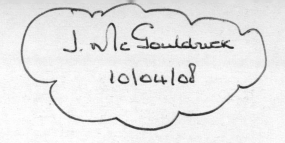

THE CLOUDSPOTTER'S GUIDE

THE CLOUDSPOTTER'S GUIDE

GAVIN PRETOR-PINNEY

AN OFFICIAL PUBLICATION OF
THE CLOUD APPRECIATION SOCIETY

www.cloudappreciationsociety.org

Chapter illustrations by Bill Sanderson

SCEPTRE

Copyright © 2006 by Gavin Pretor-Pinney

First published in Great Britain in 2006 by Hodder & Stoughton
A division of Hodder Headline

This paperback edition published in 2007

The right of Gavin Pretor-Pinney to be identified as the
Author of the Work has been asserted by him in accordance
with the Copyright, Designs and Patents Act 1988.

A Sceptre paperback

7

A CIP catalogue record for this title
is available from the British Library.

ISBN 978 0 340 89590 0

Design, typesetting and diagrams by Gavin Pretor-Pinney
in Garamond, Base Serif and Dalliance Script.

Printed and bound by Clays Ltd, St Ives plc

Hodder Headline's policy is to use papers that are natural, renewable
and recyclable products and made from wood grown in sustainable
forests. The logging and manufacturing processes are expected to
conform to the environmental regulations of the country of origin.

Hodder and Stoughton Ltd
A division of Hodder Headline
338 Euston Road
London NW1 3BH

For Liz.

Contents

☁

The Low Clouds

The Middle Clouds

The High Clouds

Not Forgetting ...

INTRODUCTION

.....

I 've always loved looking at clouds. Nothing in nature rivals their variety and drama; nothing matches their sublime, ephemeral beauty.

If a glorious sunset of Altocumulus clouds were to spread across the heavens only once in a generation, it would surely be amongst the principal legends of our time. Yet most people barely seem to notice the clouds, or see them simply as impediments to the 'perfect' summer's day, an excuse to feel 'under the weather'. Nothing could be more depressing, it seems, than to have 'a cloud on the horizon'.

A few years ago, I decided that this sorry state of affairs could not possibly be allowed to continue. The clouds deserved better than to be regarded merely as a metaphor for doom. Someone needed to stand up for clouds.

So, in 2004, I started a society devoted to doing just that. I called it The Cloud Appreciation Society and launched it during a lecture I gave at a literary festival in Cornwall. In case anyone at the talk felt moved to join the society, I'd made some official badges, and was surprised to see a rush of people come up for them at the end.

Of course, an organisation only exists when it has a website. So, a few months after the talk, I launched the society on the Internet. Initially – like the clouds themselves – membership was free, and word soon spread.

People sent in their cloud photographs, which I put up on the gallery pages for others to look at. The early trickle of submissions soon swelled to a torrent. Stunning images were arriving of rare and beautiful formations: lenticularis wave-clouds over the peaks of the Swiss Alps, rippled Cirrocumulus layers in the warm hues of the sunrise, Cumulus clouds shaped like elephants, cats, Albert Einstein and Bob Marley.

Soon, I had to start charging a nominal membership fee to cover costs, since people were joining from all over the world. They contributed cloud paintings and cloud poetry to be added to the site. I started a chat area, so that visitors would finally have a forum in which to discuss important cloud-related matters.

Some members were meteorologists and cloud physicists but most had no professional involvement with the weather at all. They ranged from octogenarian ex-glider pilots to infants of just a few months old. We all know that babies are amongst the most enthusiastic cloudspotters in the world, but I was still amazed at their ability to fill in the membership form.

The love of clouds seemed to transcend national and cultural boundaries and people joined from all across Europe, from Australia and New Zealand, from Africa, America and Iraq. By the end of the first year, we had 1,800 members, in 25 countries – all united by nothing more than an appreciation of the celestial mists.

Members soon asked me to recommend cloud books suitable for the general reader. So I looked around and decided that, save the odd glossy picture book, nothing quite fitted the bill.

Which is how *The Cloudspotter's Guide* came about. It is a guide to all the delightful and eccentric characters in the cloud family, illustrated with photographs contributed by members of The Cloud Appreciation Society. I don't present it as a meteorological text – there are already many fantastic examples of these, written by people who know a lot more than I do (and I confess to having plundered them all shamelessly). It is more serious than that – it's a celebration of the carefree, aimless and endlessly life-affirming pastime of cloudspotting.

Gavin Pretor-Pinney
London
February 2006.

The Manifesto
of The Cloud Appreciation Society

....

We believe that clouds are unjustly maligned
and that life would be immeasurably poorer without them.

☁

We think that clouds are Nature's poetry,
and the most egalitarian of her displays, since
everyone can have a fantastic view of them.

☁

We pledge to fight 'blue-sky thinking' wherever we find it.
Life would be dull if we had to look up at cloudless
monotony day after day.

☁

We seek to remind people that clouds are expressions of
the atmosphere's moods, and can be read like those
of a person's countenance.

☁

We believe that clouds are for dreamers and their contemplation
benefits the soul. Indeed, all who consider the shapes they see
within them will save on psychoanalysis bills.

☁

And so, we say to all who'll listen:

*Look up, marvel at the ephemeral beauty,
and live life with your head in the clouds.*

I am the daughter of Earth and Water,
And the nursling of the Sky:
I pass through the pores of the ocean and shores;
I change, but I cannot die.
For after the rain, when with never a stain
The pavilion of heaven is bare,
And the winds and sunbeams with their convex gleams
Build up the blue dome of air,
I silently laugh at my own cenotaph,
And out of the caverns of rain,
Like a child from the womb, like a ghost from the tomb,
I arise, and unbuild it again.

Percy Bysshe Shelley, 'The Cloud'

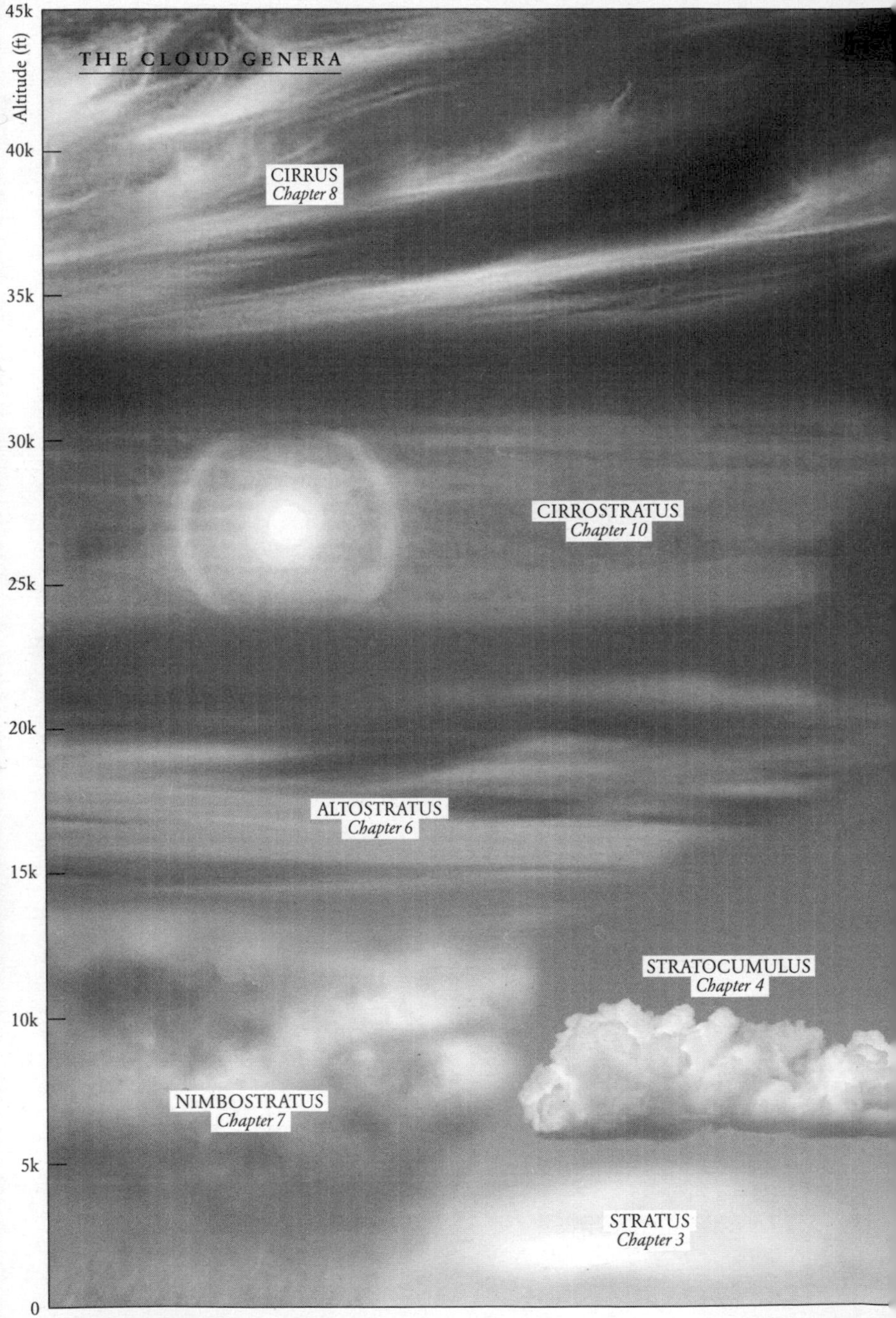

THE CLOUD GENERA

Altitude (ft)

45k

40k

CIRRUS
Chapter 8

35k

30k

CIRROSTRATUS
Chapter 10

25k

20k

ALTOSTRATUS
Chapter 6

15k

STRATOCUMULUS
Chapter 4

10k

NIMBOSTRATUS
Chapter 7

5k

STRATUS
Chapter 3

0

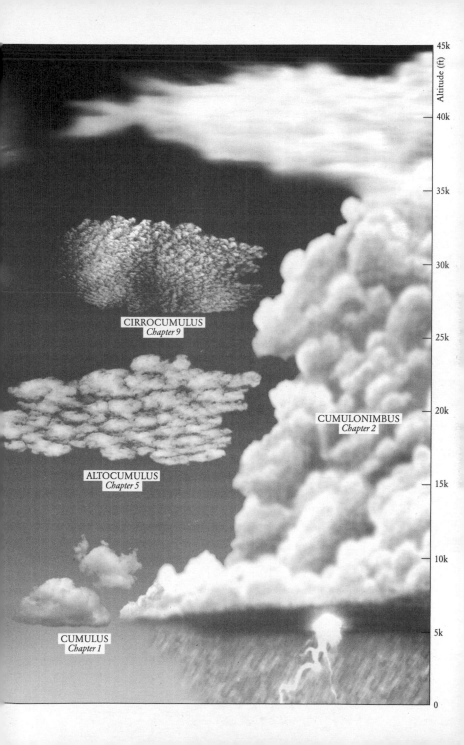

Altitude (ft)

45k

40k

35k

30k

CIRROCUMULUS
Chapter 9

25k

CUMULONIMBUS
Chapter 2

20k

ALTOCUMULUS
Chapter 5

15k

10k

CUMULUS
Chapter 1

5k

0

CLOUD CLASSIFICATION TABLE

Clouds are classified according to a Latin 'Linnean' system (similar to the one used for plants and animals), which is based on their heights and appearance. Most clouds fall into one of ten basic groups, known as 'genera'. They can further be defined as one of the possible 'species' for that genus, and any combination of the possible 'varieties'. There are also various accessory clouds and supplementary features that sometimes appear in conjunction with the main cloud types.

(If all this Latin freaks you out, don't worry – it freaks me out too.)

	GENUS	SPECIES (CAN ONLY BE ONE)	VARIETIES (CAN BE MORE THAN ONE)	ACCESSORY CLOUDS AND SUPPLEMENTARY FEATURES	
LOW CLOUDS	Cumulus	humilis		pileus	arcus
		mediocris	radiatus	velum	pannus
		congestus		virga	tuba
		fractus		praecipitatio	
	Cumulonimbus *(extends through all three levels)*			praecipitatio	pileus
		calvus		virga	velum
		capillatus	(none)	pannus	arcus
				incus	tuba
				mamma	
	Stratus	nebulosus	opacus		
		fractus	translucidus	praecipitatio	
			undulatus		
	Stratocumulus		translucidus		
			perlucidus		
		stratiformis	opacus	mamma	
		lenticularis	duplicatus	virga	
		castellanus	undulatus	praecipitatio	
			radiatus		
			lacunosus		
MIDDLE CLOUDS	Altocumulus		translucidus		
		stratiformis	perlucidus		
		lenticularis	opacus	virga	
		castellanus	duplicatus	mamma	
		floccus	undulatus		
			radiatus		
			lacunosus		
	Altostratus		translucidus	virga	
			opacus	praecipitatio	
		(none)	duplicatus	pannus	
			undulatus	mamma	
			radiatus		
	Nimbostratus *(extends through more than one level)*			praecipitatio	
		(none)	(none)	virga	
				pannus	
HIGH CLOUDS	Cirrus	fibratus	intortus		
		uncinus	radiatus		
		spissatus	vertebratus	mamma	
		castellanus	duplicatus		
		floccus			
	Cirrocumulus	stratiformis			
		lenticularis	undulatus	virga	
		castellanus	lacunosus	mamma	
		floccus			
	Cirrostratus	fibratus	duplicatus	(none)	
		nebulosus	undulatus		

The Low Clouds

ONE

CUMULUS

The cotton wool tufts that form
on a sunny day

Leonardo da Vinci once described clouds as 'bodies without surface', and you can see what he meant. They are ghostlike, ephemeral, nebulous: you can see their shapes, yet it's hard to say where their forms begin and end.

But the Cumulus cloud is one that challenges da Vinci's description. Rising in brilliant-white cauliflower mounds, it looks more solid and crisply defined than other cloud types. As a child, I was convinced that men with long ladders harvested cotton wool from these clouds. They look as if you could just reach up and touch them – and, if you did, they would feel like the softest things imaginable. The most familiar and 'tangible' of the cloud family, this is a good type for budding cloudspotters to cut their teeth on.

Cumulus is the Latin word for 'heap', which is simply to say that these clouds have a clumpy, stacked shape. The people who concern themselves with such things divide them into humilis, mediocris and congestus formations – these are known as 'species' of Cumulus. Humilis, meaning humble in Latin, are the smallest, being wider than they are tall; mediocris are as tall as they are wide, and congestus are taller still.

It is the smaller ones that generally start forming over land on sunny mornings. And because neither they nor their mediocris brothers produce any precipitation, they are widely recognised as 'fair-weather clouds' – a pair of puffy fingers up at those who can

HOW TO SPOT
CUMULUS CLOUDS

Cumulus are low, detached, puffy clouds that develop vertically in rising mounds, domes or towers, and have generally flat bases. Their upper parts often resemble cauliflowers and they appear brilliant white when reflecting high sunlight, but can look dark when the Sun is behind them. Cumulus tend to be randomly scattered across the sky.

TYPICAL ALTITUDES*: 2,000–3,000ft

WHERE THEY FORM: Worldwide, except in Antarctica (the ground is too cold for thermals).

PRECIPITATION (REACHING GROUND): Generally none, except for brief showers from congestus.

Cumulus humilis

Cumulus mediocris

Cumulus congestus

CUMULUS SPECIES:

HUMILIS: Minimal vertical extent. They look flattened and appear wider than they are tall. Do not cause precipitation.

MEDIOCRIS: Moderate vertical extent. Might show protuberances and sproutings at the top. Appear as tall as they are wide. Do not cause precipitation.

CONGESTUS: Maximum vertical extent. The tops are like cauliflowers. Appear taller than they are wide. Cause brief downpours.

FRACTUS: Ragged edges and broken up. Can form in the moist air below rain clouds.

CUMULUS VARIETIES:

RADIATUS: When Cumulus have formed into rows, or 'cloud streets', which are roughly parallel to the wind direction. Due to perspective, the rows appear to converge towards the horizon.

Cumulus mediocris radiatus

NOT TO BE CONFUSED WITH...

STRATOCUMULUS: Cumulus clouds are detached, not joined into a layer like Stratocumulus.

ALTOCUMULUS: Cumulus are not usually as regularly spaced as a layer of the higher Altocumulus. The clouds also look larger than the clumps of the Altocumulus. When they are above the cloudspotter, Cumulus appear larger than the width of three fingers, held at arm's length.

CUMULONIMBUS: which often develops from a large Cumulus congestus. A cloud is still a Cumulus when its upper region has a sharp outline, compared with the softer top of the Cumulonimbus.

* These approximate altitudes (above the surface) are for mid-latitude regions.

right: Michael Rubin (member 329), Bottom: Paul Cooper (member 1523)

only think of clouds as the opposite of fine weather. A lazy sunny afternoon beneath the drifting candyfloss curls of the *Candyfloss clouds* Cumulus is far finer than the flat monotony of a cloudless *clouds* sky. Don't be brainwashed by the Sun fascists – fair-weather Cumulus have a starring role in the perfect summer's day.

There is one other species of this cloud: Cumulus fractus. This has a much less puffy shape, its edges being fainter and more ragged. It is the way a Cumulus looks when it is decaying at the ripe old age of ten minutes or so.

Besides being divided into species, each of the ten main cloud types – each 'genus' of cloud – has a number of possible 'varieties'. These are characteristics of appearance that are often observed in that cloud type. For the Cumulus cloud, the only recognised variety is Cumulus radiatus, which is when the clouds are lined up in files parallel to the wind. These rows of cotton wool tufts are sometimes called 'cloud streets'.

Although Cumulus is generally associated with fine weather, any cloud can under certain conditions develop into a rain-bearing formation, and Cumulus is no exception. The innocuous Cumulus humilis and mediocris can on occasions grow into the angry, towering Cumulus congestus, which it must be said is anything but a fair-weather cloud. It is well on the way to becoming the enormous awe-inspiring Cumulonimbus thundercloud, and can itself produce moderate to heavy showers. Whilst the development of Cumulus clouds from humilis right up to the congestus and beyond can be a daily occurrence in the hot, humid tropics, it is less common in temperate climes. Nevertheless, if you see Cumulus develop into the tall congestus stage before midday, there is a distinct possibility of heavy showers by the afternoon. Attention all cloudspotters: 'In the morning mountains, in the afternoon fountains.'

☁

THE DISTINCTIVE SHAPES of Cumulus clouds may go some way to explaining why they are the cloud of choice in the drawings of young children. No six-year-old's picture of a family in front of

Laurette Saris (member 1593)

Be careful when you complain about the clouds – they can hear what you are saying. Cumulus, in particular, give as good as they get.

their house feels complete without a few puffy Cumulus floating in the sky above. Children just have a fascination with clouds. Can it be that, wheeled around in prams staring up at the sky as infants, they develop a deep connection with the clouds – like young chicks forming a familial bond with the first thing they see out of the egg? Who knows? Their drawings may show people with arms coming out of their necks, whose eyes aren't even attached to their faces, but young children seem to capture the organic shapes of Cumulus clouds pretty well. No doubt they are just easier to draw than other types. Perhaps, however, their ubiquity in junior school drawings is due to something more fundamental than this.

Cumulus also feel like the most generic and basic of all the varieties. Picture a cloud in your mind and it is likely to have the *The cloud* shape of a Cumulus, which is probably why it was their *of choice for* gentle, bulging curves that were used by Mark Allen, a 22- *six-year-olds* year-old graphic designer, when he created the icons for the BBC weather forecasts in 1975. Back then they were in the form of rubber-coated magnets, which the TV forecasters would slap on to the map of Britain. I'd snigger, along with the rest of the nation, each time these fell to the ground after they turned their backs.

The Cumulus symbols were used to stand for cloud cover for thirty years until 2005, when the BBC weather graphics were completely redesigned into a dynamic 3D system, which showed how cloud coverage and rainfall distribution varied in real time. Whilst the new system gave a much more accurate indication of cloud cover, viewers complained that the way the camera panned and swooped across the computer-generated map made them feel sick. But

Being the best cloudspotters in the world, six-year-olds rarely miss the opportunity to bung some Cumulus into their drawings.

perhaps that was just an excuse and, like me, they were merely sad to have to say goodbye to the friendly Cumulus symbols.

⌒

ALTHOUGH CLOUDSPOTTING is an activity best undertaken with time on your hands, it is something that everyone can enjoy. Clouds are the most egalitarian of Nature's displays, since each one of us has a good view of them, so it really doesn't matter where you are. A little elevation never goes amiss, of course, but this could as easily be provided by a tower block as a mountain range of outstanding natural beauty. More important is the frame of mind you are in while cloudspotting. You are not a trainspotter, so standing on a hill with a notebook and pen poised to tick off the different types will end in disappointment. So will any attempt to write down their serial numbers.

The now-defunct cloud symbol, designed for the BBC weather forecasts by Mark Allen when he was considerably older than six.

A cloudspotter is not a cataloguer – meteorologists are busy indexing the different genera, species and varieties of clouds on your behalf. They call it work. Yours is a far more gentle and reflective pursuit – one that will lead to a deeper understanding of the physical, emotional and spiritual world. John Constable, perhaps Britain's best cloud painter, saw the sky as 'the keynote' and 'the chief organ of sentiment' in his landscape paintings. And his cloudscapes do have a drama and vitality to them that I feel is lacking in the rural idylls below.

Constable believed that 'we see nothing truly till we understand it'. I agree. If cloudspotters appreciate the way clouds form, what makes them look the way they do, how they shift from one formation to the next, how they grow and develop, how they decay and dissipate, they will have learned more than mere principles of meteorology. René Descartes, the French Jesuit philosopher of the seventeenth century, wrote of clouds:

> *Since one must turn his eyes toward heaven to look at them, we think*
> *of them… as the throne of God… That makes me hope that if I can*
> *explain their nature… one will easily believe that it is possible in some*
> *manner to find the causes of everything wonderful about Earth.*[1]

☁

SO WHAT EXACTLY is a Cumulus cloud? It may feel rather unsatisfying to hear that it is just water. And yet, like all clouds, that is all it is. The curious cloudspotter might therefore wonder why it looks so different from a glass of the stuff down here on the ground. The cloud's white, opaque appearance is because the water

It's only water! is in the form of countless tiny droplets (well, around 10,000,000,000 per cubic metre in actual fact), each only a few thousandths of a millimetre across. And this array of innumerable tiny surfaces scatters the light in all directions, giving the cloud its diffuse, milky appearance as compared with the single surface of a container of water. It is like the rough face of etched glass compared with a smooth pane: all the minute angled surfaces of the roughened glass make it look white as they scatter the light every which way.

According to ancient Hindu and Buddhist beliefs Cumulus clouds are the spiritual cousins of elephants, which is why the animals are worshipped, with a view to bringing rain after India's scorching summer heat. 'Megha', meaning cloud in classical Hindi, is the name used to address elephants in these prayers. The Sanskrit

Cumulus mediocris – as tall as it is wide.

creation myths describe how elephants created at the beginning of time were white, had wings to fly, could change their shape at will and had the power to bring rain. Although they have now lost these magical powers, the present-day descendants of those early Über-elephants are still believed to have an affinity with the clouds – especially the albino ones.

It is somewhat alarming to learn that eighty elephants weigh about as much as the water droplets in a medium-sized Cumulus –

27

The droplets in a typical Cumulus weigh the same as eighty elephants.
But this one looks like it only weighs as much as one baby elephant.

a Cumulus mediocris – would if you added them all together.* For, though the droplets in a Cumulus cloud are extremely small, there are one hell of a lot of them. Given that elephants don't tend to fly these days, how exactly does the water equivalent of eighty of them rise to form a Cumulus?

There's a clue in the cloud's tendency to appear on a sunny day. For when the Sun is shining, currents of air known as thermals, or convection currents, start to form as it warms the ground. These rising plumes of air are the light turbulence you feel as you pass through Cumulus in an aeroplane. They are the reason why hang gliders and eagles will head towards this type of cloud, knowing that it is a celestial signpost for the updraughts that give them lift. Thermals are the invisible spirits that give life to the Cumulus. They bring it into being, flowing through it, animating it.

The spirit of a Cumulus

---

* This is assuming the cloud occupies one cubic kilometre (about 0.24 cubic miles), which is not particularly large for a Cumulus. The droplets will commonly have a combined weight of 200,000kg. The average Asian elephant weighs 2,500kg.

To understand the formation of thermal convection currents is to glimpse the soul of a Cumulus cloud. They are what get the moisture up there in the first place and also what help the cloud's droplets to stay airborne for the ten minutes or so of a typical Cumulus's life.

It's a lot like the movement of the blobs of oil in a lava lamp. The mixture of oil and coloured water inside the lamp moves upwards by the same process of convection as air on a sunny day. Although the lamp contains liquids rather than gases, the principle is the same.

The oil in the lamp is normally just a bit denser than the water, and so sits at the bottom, but when the bulb in the base warms it up, the oil expands, becomes a little less dense, and begins to float up lazily through the water. The air outside behaves in a similar way. A ploughed field that has been warmed by the Sun can act like the bulb in the lamp, warming the air above – making it expand, become less dense and float upwards through the surrounding cooler air. The invisible moisture carried in the rising thermal is what can end up as Cumulus, or, in the words of the American poet Maria White Lowell, 'little tender sheep, pastured in fields of blue… with new-shorn fleeces white'.

③ Away from the lamp, the oil cools and contracts again, starting to sink back down. *Cooler air away from the ground on a sunny day also descends to take the place of air that has risen in a thermal.*

② The warmed oil expands, becomes less dense than the water, and floats up. The movement is called convection. *The air above a Sun-warmed field also expands and floats upwards.*

① The hot bulb warms the oil at the bottom of the lamp. *In the same way a field, heated up by the Sun, warms the air above it.*

Whilst it doesn't look much like a field on a summer's morning, unless you have been at the jazz cigarettes, the lava lamp does show how air can rise in a thermal, which carries water vapour up to form a cumulus cloud.

Remember that Cumulus are individual clouds, quite different from the large layers you see in an overcast sky. For it so happens that some surfaces absorb and give off the Sun's heat better than others, and thus a pocket of air will rise by convection more readily over here than one over there. Tarmac, for instance, will heat the air more efficiently than a grass field. A hillside facing the Sun will do so faster than one in shadow. Cloudspotters will be pleased to see this most clearly demonstrated when sailing around a small island on a sunny day. The surface of the island is warmed by the Sun's radiation more readily than the sea around it, and a puffy, white Cumulus cloud can often be seen poised above it, fed by the thermal coming off the ground. South Sea Islanders would use Cumulus clouds as beacons, navigating towards an atoll well before the land itself became visible.

Since they form on top of these independent convection currents, Cumulus are separate, individual clouds. This is one of the main ways in which they differ in appearance from other cloud types. Each is the visible summit of a towering transparent column of air – like a bright white toupee on a huge invisible man. And the Cumulus can soon drift off from its thermal host – the wig plucked from his head, swirling and folding in on itself in slow motion as it is swept along in the breeze.

ONE THING THAT MAKES the Cumulus such an appealing cloud to the fledgling cloudspotter – besides, of course, its fair-weather associations – is that it looks so damn comfortable. Who hasn't gazed up and dreamt of falling asleep in the Cumulus's plump, white folds? These clouds are like furniture fit for gods. Which must be why they have historically appeared in religious imagery as the sofas of the saints. Whereas up until medieval times God was only ever depicted in Western art as a hand or an eye emerging from the clouds, from the Early Renaissance onwards clouds were commonly used in religious paintings as supports for the deities.

I recently spent seven months living in Rome, and though the summer skies were often entirely devoid of clouds, I soon noticed
Rome – a that down at ground level the city had more than its fair
cloudy city share. It struck me that, on every street corner, the Baroque church interiors were frescoed with billowing white Cumulus cushions, upon which sat the apostles and saints looking down to the congregation below. Bernini's famous sculpture, *St Teresa in Ecstasy*, shows the saint falling back into the folds of a Cumulus

Paulo Uliana (member 1598)

The humble Cumulus humilis – never hurt a soul.

cloud hewn from travertine stone. And Raphael and Titian's Renaissance paintings in the Vatican, from a hundred years before, depict the Madonna holding the baby Jesus in her arms, or being crowned at her Ascension, always suspended on a generous bed of vapours. Even the mosaic in the Basilica of Saints Cosma and Damiano next to the Roman forum, dating back to the sixth century, shows Jesus dressed in a toga standing on a carpet of clouds, tinged red and orange with the hues of the setting Sun.

Being half-way between heaven and earth, clouds were the perfect religious symbols for separating the divine and mortal realms – vaporous furniture allowing artists to place everyday mortals and divine beings together in the same image. For so many artists producing Christian imagery, ample, virginal Cumulus clouds were what divided all that was heavenly and pure from the sinful, mortal realm below.

The clouds and Christianity have always gone hand in hand. Biblical references are plentiful. In Exodus, God appeared on Mount Sinai in a cloud, which at once concealed and revealed him, and he led the ransomed Israelites across the desert in a pillar of

A Roman Catholic prayer card from 1892 shows Jesus, Joseph and the Virgin Mary sitting on Cumulus clouds. These make them look rather comfy above the poor souls in the flames of Purgatory.

cloud – the glory cloud – preceding them as they marched, stopping when it was time to make camp, rising up when it was time to move on. In the Book of Acts, Jesus was received into a cloud as he ascended to heaven after his resurrection, whilst *The Passing of Mary, Second Latin Form* describes the Virgin rising to heaven on a cloud, after the apostles had been transported to her deathbed on them. In Judaic mythology 'Bar Nifli', or 'Son of a Cloud', is a title for the Messiah. According to the Book of Daniel, he will appear riding a white cloud.

But associations between clouds and the divine are by no means limited to Christo–Judaic faiths: Islamic esotericism holds that Allah was in the state of a cloud before his manifestation; the Japanese god of thunder and lightning, Raiden, prevented the Mongols from invading Japan in 1274 by sitting on a cloud and showering lightning arrows upon their fleet; Sun Wu-Kung, the Monkey King who brought the old scriptures from Buddha in the Western Heaven back to China in the classic tale *Journey to the West*, got around by 'cloud dancing', which allowed him to cover huge distances by leaping from cloud to cloud.

And the list goes on: Parianya, meaning 'rain cloud', is the old Indian god of rain and vegetation, who is married to the fertilised Earth and represented in the form of a bull; Perkons, one of the main deities in Baltic folk religion, is the thunder god, whose main role is to bring fertility; Ngai, the creator god and chief deity of the Masai tribes of Kenya and Tanzania, appears as a red cloud when he is angry and a black one when he's in a good mood. Wondjina, primal beings of Aboriginal myth, are cloud and rain spirits who descended into caves in the Dreamtime, and one of whom rose to the sky and formed the Milky Way… I could go on, but I think you get the idea.

As children, we look up to our parents – *physically*, we look up towards them – and are parents not the nearest things to gods for a young child? Perhaps it is why, as adults, we look up to our gods in the sky. No doubt, it is also because rain and sunshine, upon which our survival has always depended, come from above and are beyond the control of earthbound mortals. Whatever the reasons, we've seen the clouds when we've looked to our deities and so have naturally associated the two. How sad that we can now go up in aeroplanes and see that there are no gods upon the clouds.

By the mid-nineteenth century, once balloon travel had become a relatively commonplace means of taking man up to cloud level, the Victorian critic and essayist John Ruskin wrote:

> *Whereas the medieval never painted a cloud but with the purpose of placing an angel upon it…, we have no belief that the clouds contain more than so many inches of rain or hail.*[2]

Mike Matthews (member 792)

Low clouds can line up parallel to the wind to form Cumulus radiatus. Also known as 'cloud streets', they're the Roman roads of the cloud world.

FOR THE FIRST two hundred years of balloon flights, hydrogen – a gas lighter than air – was the most common means of lift. The dangers of using such a highly combustible gas and the lack of easy control of altitude meant that by the 1960s manned balloon flights were more commonly powered by heating the air in the balloon. This makes it expand and become less dense than the air around and float upwards. It is the very same principle as the rising of thermals on a sunny day. In fact, Cumulus clouds can form above fires. Known as pyrocumulus, these appear as Cumulus mounds atop the plumes of smoke from stubble burning or wildfires. The heat of the flames induces thermal currents of air, which can carry moisture up with them.

But what does it mean for the air in a thermal to 'contain' moisture as it rises? And how can this invisible moisture magically change into the visible droplets that make up a Cumulus cloud? It might seem rather less mysterious if you remember that you are creating clouds when you see your breath on a cold day. I remember being fascinated by this when my father took me to the park on freezing autumn mornings to collect conkers. I'd wave my mittened hands through the misty breath, both delighted by it and disappointed that, unlike the clouds above, the ones I created vanished so quickly. I wished they would form a row of 'breathprints' leading all the way back to the car. They may not have stuck around – instead just evaporating away into the morning air – but they were indeed real clouds. Scale and altitude aside, they were in fact no different from Cumulus clouds.

The air we breathe out is always packed with water vapour. Our bodies make sure that it is, since our moist bronchial tubes are designed to stop dust and pollution getting into our lungs. At least four per cent of the weight of the warm air we breathe out consists of individual water molecules, which are colliding with anything and everything in their path – just like the oxygen, nitrogen and other molecules that make up air. When in the form of molecules like this, water is a gas and is called 'water vapour'. However, individual molecules are much too small for us to see, and so air,

Steve Maniscalco (member 1397)

Pyrocumulus can form in the hot, moist air that rises from forest fires.

regardless of how much water vapour it contains, is transparent. Only when the water molecules stick together into clumps are we able to see them.

And this is exactly what happens when we exhale on a cold day. Our warm, moist breath mixes with the cold air and quickly drops in temperature. As any gas cools, its molecules slow down. And when this happens to the water molecules in our breath, they more readily stick together.

Likewise, thermals rising from the ground carry water vapour up with them and, if they are cool enough when they rise, their decelerating water molecules also become more likely to stick together and some of them form into the countless tiny droplets that make up a Cumulus cloud.

THE CLASSICAL GREEK GOD Zeus was the lord of the sky, the god of rain and the cloud-gatherer. He and his wife, Hera, had a somewhat tempestuous relationship, mostly because he was such a philanderer. Hera was jealous of his lovers, and Zeus sought vengeance on anyone who took a shine to her. Ixion was one such

Jupiter and Io, 1531, by Correggio – the seedy world of sixteenth-century cloud pornography.

character who had the cheek to pursue Hera after he had been invited to Olympus by Zeus himself. Having heard that something fishy was going on, Zeus decided to test Ixion's intentions and fashioned a cloud that resembled his wife. Ixion had it away with the cloud and was killed by Zeus. The cloud later gave birth to Centaurus who, like his unruly offspring the centaurs, was part man and part horse. I guess this is a warning to over-eager cloud-spotters: don't get too intimate with a cloud.

Zeus himself seems to have had a bit of a fetish for clouds – if *Jupiter and Io*, a sixteenth-century painting by the Italian Renaissance artist Correggio, is anything to go by. It shows a naked woman in the throes of sexual abandon as she is ravaged by a dark Cumulus cloud.

The painting was part of a triptych that Correggio produced in the 1530s, depicting the Loves of Jupiter, the Roman name for the Greek god Zeus. Correggio painted the god making love to Io, who was Hera's priestess. In the myth, as recounted by the Roman poet Ovid, the over-sexed Jupiter lusts after Io and tries to take advantage of her in the meadows of Lerna. So concerned is he that his wife will see him philandering with this mortal bit of fluff that he hides himself in a dark cloud. And as Ovid explains, Io does her best to get away:

For while he spoke she fled,
and swiftly left behind the pasture fields
of Lerna, and Lyrcea's arbours, where
the trees are planted thickly. But the God
called forth a heavy shadow which involved
the wide extended earth, and stopped her flight
and ravished in that cloud her chastity.[3]

In contrast to the violent scene described by Ovid, Correggio chose to depict Io looking as if she was having a rather good time with the cloud, and in the process created what must be the most erotic painting of a cloud ever put to canvas: the first – and sadly the last – example of sixteenth-century cloud pornography.

The Cumulus with which Correggio shrouded the randy Jupiter was dark blue-grey in colour. The darkness of a Cumulus cloud depends, firstly, on whether you are looking at the side in shadow and, secondly, on the brightness of the sky or other clouds behind it. But it also depends on the number of water droplets that the cloud contains, for it is these that scatter the sunlight and prevent some of it from passing through. The more laden with *A dark,* droplets a cloud is, the darker it will appear with the Sun *randy* behind. Cloudspotters will note that as a Cumulus grows in *Cumulus* size from its small humilis form, through the mediocris stage into a towering thick congestus, its base will appear darker and darker as the thickening cloud blocks out more and more sunlight. How appropriate, then, for Correggio to have painted the over-sexed god, brimming with evil intent and lust for the beautiful Io, as a dark grey Cumulus congestus – a phallic tower of a cloud, so laden with water that it is bursting to rain down its load.

☁

WITH ITS CHURNING INNARDS stirred by convection currents, the Cumulus can be a skittish and unpredictable cloud. Viewed from the ground, its internal movement seems gentle, even lazy, but remember that when an object is some way off it seems to move more slowly (a jet aeroplane, high above, can appear to move at a snail's pace). In reality, the turbulence within the cloud can be brisk. And once it starts growing, an innocuous fair-weather humilis can build in a matter of hours to an enormous congestus, whose darkening base warns of sudden, heavy showers.

How does a Cumulus humilis manage to grow and build like this? It's a question on every cloudspotter's lips. If a Cumulus, forming on a thermal that rises from a source on the ground, is then blown away by the wind (as so often happens when low clouds drift across our skies) then why should the air within it keep rising, higher and higher, and build it into a billowing tower? With the thermal left behind, what is the source of the lift? This is where the lava lamp analogy falls down, for its blobs of oil, cooling as they rise from the lamp, soon contract and sink again. Why doesn't the air in a building Cumulus just do the same?

It is because of something called 'latent heat'. Put aside your prejudices towards anything that sounds like a physics lesson and listen up, for this is essential to understanding the playful, carefree behaviour of the fair-weather Cumulus. Remember Constable's sentiment: 'we see nothing truly till we understand it'. Of course, anyone can appreciate a glorious cloudscape, but it strikes me that the more a cloudspotter understands about clouds' behaviour, the more their beauty opens up.

Latent heat of condensation is the warmth that is given off when free-flying water molecules join together into droplets of liquid, as they do when a Cumulus forms at the top of a rising thermal. It so happens that when water condenses from vapour into liquid droplets it gives off heat into the surrounding air.

It might be easier to understand this principle by considering the opposite – when liquid water evaporates into a vapour and it *The sweaty* takes up heat from its surroundings. If this makes things any *business of* clearer, consider the sweat forming on my brow when I go *latent heat* out for a jog on a summer afternoon. (This is fictional sweat, on my fictional brow, on a fictional summer afternoon, for jogging is not something that I generally do.)

The sweat evaporates from my forehead in the breeze and its water molecules carry heat away with them, leaving my forehead a little cooler (and stopping me from overheating). In fact, it is not

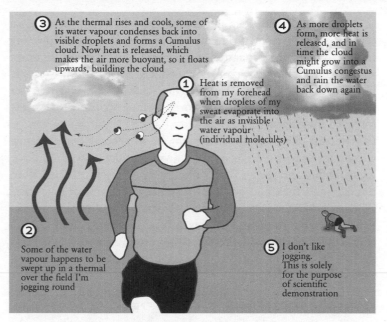

How a Cumulus can grow when I'm out jogging.

inconceivable that the molecules of my sweat might get swept up in a thermal forming over the field that I am running around – that they might ascend with the curling, twisting current of air, cooling with the air as they rise.

They might reach a height where the air has cooled enough for many of them to start joining back together into clumps – into cloud droplets. The heat that the molecules of my sweat took away with them as they evaporated from my forehead is released again when they form into the cloud. This heat, released as water vapour condenses, is what scientists call latent heat, and it is central to how little Cumulus clouds grow into big Cumulus clouds.

When the droplets form and release their latent heat, the surrounding air warms up a bit. This makes the air expand more, become more buoyant and can cause it to float upwards with renewed vigour. The latent heat released within the Cumulus is how the cloud grows vertically. It gives the air added lift and is why the Cumulus has puffy mounds at its top.

In this way, the Cumulus can build, seemingly of its own accord, from humilis to mediocris and even to the towering congestus. And the congestus cloud might rain its moisture down again. And I might happen to be jogging back around the field below as it does so. And the water molecules that had started out as my sweat might fall back down in a raindrop and land on my forehead once again.

This sort of circular futility is why I don't generally go jogging.

☁

CLOUDS MAY BLOCK OUT our beloved Sun, the very heat of which gives rise to them, but they also let us see it. A thin layer of cloud called a Stratus is the very thing that can allow us to look directly at the Sun without damaging our eyes. By obscuring, clouds let us see the light.

The contradiction was beautifully articulated by the medieval Christian mystic who wrote *The Cloud of Unknowing*. As befits the author of a title like that, his identity is a mystery. Scholars assume he was some sort of priest or monk, but no one can be sure. It seems that this mystery figure wrote *The Cloud of Unknowing* sometime around the 1370s, but even that is a matter of dispute. He used the image of a cloud to describe the impossibility of any mortal being able to get to know God.

Clouds and medieval monks

The author – let's call him 'Monk X' – was an 'apophatic' mystic, which means he was of the opinion that no matter how devout a Christian you are, you can never actually conceive of what God is like, regardless of how hard you think about it. The powers of reason can tell us more about what God is *not* than about what He *is*. A Christian may well be able to contemplate the words of God through studying the Bible, and he may indeed be able to communicate with God through prayer, but Monk X claimed that he'll never be able to make sense of what God is. He will forever be separated from Him by a cloud of unknowing.

The sooner the Christian gets used to this inevitable impediment to being able to 'see' God, X argues, the better:

> *This darkness and this cloud is, howsoever thou dost, betwixt thee*
> *and thy God, and letteth thee that thou mayest neither see Him*
> *clearly by light of understanding in thy reason, nor feel Him in*
> *sweetness of love in thine affection.*
> *And therefore shape thee to bide in this darkness as long as thou*
> *mayest, evermore crying after Him that thou lovest. For if ever thou*
> *shalt feel Him or see Him, as it may be here, it behoveth always to*
> *be in this cloud in this darkness.*[4]

He didn't say that Christians should therefore just give up dialogue with their God. Only that this inevitable cloud of unknowing means the intellect and reason will never reveal His true nature. The cloud is a sufferance for the Christian – an impediment, an obstruction – which, like a real cloud blocking out the Sun, keeps him away from his God. Best to accept it, rather than fight it, advised Monk X. Only then is there the possibility of knowing God through ways other than rational thought.

The Christian should let himself be drawn up into this cloud. Indeed, he should put a 'cloud of forgetting' between himself and his rational mind. He should forget, forget, forget. Only then can he begin to let the 'blind stirrings of love' develop, through which he will come to 'know' God:

> *And therefore lift up thy love to that cloud: rather, if I shall say thee*
> *sooth, let God draw thy love up to that cloud and strive thou through*
> *help of His grace to forget all other thing.*[5]

To be up in the cloud of unknowing is to be nearer to God – nearer than those who struggle in vain to work Him out. Monk X was telling Christians to accept the limitations of their understanding and know God by unknowing.

If you think this all sounds rather Zen-like for a Christian, then you won't be the first. But the apophatic mystics were working in an entirely different religious system from their Oriental counterparts. For Monk X, the idea of a cloud separating us from the divine light of God was, in spite of all its frustrations, a profound expression of Christian faith.

Cumulus congestus is the largest of the species and often develops into
a Cumulonimbus thundercloud.

FLEDGLING CLOUDSPOTTERS couldn't be expected to be *au fait* with the temperature gradient of the lower atmosphere at any moment. Nor would we want them to be. So it may appear rather random as to whether a Cumulus humilis develops into a towering Cumulus congestus or not. When it does, the flare-up can feel like an argument between lovers, where a light-hearted remark builds into a full-blown row without either seeing it coming.

All that is needed is some lingering tension in the air between them. That, and for neither to have the wherewithal to put a lid on it. And if there is enough moisture in the lower atmosphere, if the action of the Sun is strong enough to form sizeable thermals and if there is no layer of warmer air up above the cloud level to put a lid on the convection currents, then the innocuous Cumulus humilis can soon build just as unpredictably through its mediocris stage into the angry head of steam of the congestus.

One thing that can put a lid on a Cumulus is what meteorologists call a 'temperature inversion'. This is a layer of air, within which the temperature increases with height, and it tends to halt the vertical growth of a cloud. When the air above a Cumulus happens to be like this, it is unable to build upwards since its warm rising convection current reaches a point where it is no longer warmer and lighter than the air around, and so it stops. If an inversion is present, the Cumulus cloud can then be forced to spread out sideways, rubbing its cotton wool shoulders against those of its neighbours, merging with them into a puffy layer that covers the sky.

Which is better: for arguments between lovers to flare out of the blue into furious rows that clear the air, or for them to be kept under control and for tensions to spread into a silent, lingering impasse? Who's to say, but as anyone familiar with *The Jerry Springer Show* knows, the flare-ups are more fun to watch. The Cumulus doesn't necessarily stop developing at its towering, congestus stage.

Cumulus clouds and Jerry Springer

Given the right conditions, it can keep on growing: building and building from its low Cumulus base to over 40,000ft up into the atmosphere – perhaps nearer 60,000ft in the tropics. Darker and darker, angrier and angrier it becomes, until it is no longer classified as a Cumulus at all, and has become the mighty thunderhead called a Cumulonimbus. This is when what started as a light-hearted tiff between lovers has well and truly got out of hand and the Jerry Springer bouncers have to come racing in from the wings.

CUMULONIMBUS

*The towering thunderclouds that
scare us senseless*

The clouds are our fluffy friends – except, perhaps, for one: the Cumulonimbus. When it comes to extreme and destructive weather, you can be sure that a Cumulonimbus will be in the thick of the action. With torrential downpours, hail storms, snowstorms, lightning, gales, tornadoes and hurricanes the enormous thundercloud can lead to untold loss of life and damage to property. It has also been known to frighten little children with its thunder.

When it is mature, this cloud can be considerably taller than Mount Everest. The largest examples tend to occur in the tropics, where they can extend from low bases, 2,000ft above the ground, up to summits at 60,000ft or so. The energy contained within a cloud like this has been estimated to be the equivalent of ten Hiroshima-sized bombs. No wonder it is often referred to as the King of Clouds.

That, however, sounds like a rather benevolent moniker to me. I prefer to think of Cumulonimbus as the Darth Vader of the cloud world and, just like the *Star Wars* villain, it is the most exciting character of the lot. With all its malevolent Force, it makes its Luke Skywalker son – the fair-weather Cumulus – look like a bit of a sissy. There's no lying on your back and dreaming of fluffy sheep when this beast is around. If it *is* the King of Clouds, then the Cumulonimbus is certainly a very angry king.

CUMULONIMBUS CLOUDS

Cumulonimbus are thunderstorm clouds, characterised by their enormous height. They are typically tall enough to reach the top of the troposphere, where they spread out in plumes of ice particles that can appear smooth, fibrous or striated. They have dark bases and produce heavy showers – often of hail – which can be accompanied by thunder and lightning.

TYPICAL ALTITUDES*:
2,000–45,000ft
WHERE THEY FORM:
Common in tropical and temperate regions. Rare in polar ones.
PRECIPITATION (REACHING GROUND):
Heavy downpours, often of hail.

Cumulonimbus calvus (means 'bald')

Cumulonimbus capillatus (means 'hairy')

CUMULONIMBUS SPECIES:
The two species are distinguished by the appearance of the cloud's top.
CALVUS: When the upper region is of soft indistinct flattened mounds, without any fibrous or striated appearance.
CAPILLATUS: When the upper region is cirrus-like and fibrous or striated, often in the shape of an anvil, plume or a disorderly mass of white hair.

CUMULONIMBUS VARIETIES:
There are no official varieties.

NOT TO BE CONFUSED WITH...
NIMBOSTRATUS: which is a dark, ragged precipitating layer, covering the sky. It can look similar to a Cumulonimbus that is directly over-head (and also appears to cover much of the sky) but the precipitation will tend to be more steady and more persistent than the short heavy showers of the Cumulonimbus. If thunder, lightning or hail is present, then the cloud is a Cumulonimbus.
CUMULUS CONGESTUS: from which a Cumulonimbus often develops. Seen from a distance, the cloud is said to have changed into a Cumulonimbus when parts of its upper region begin to lose their sharp edges, due to the droplets freezing into ice crystals. Thunder, lightning or hail will also identify the Cumulonimbus.

* These approximate altitudes (above the surface) are for mid-latitude regions.

IT IS ALSO A SERIOUS hazard to aircraft. This cloud's hail can grow large enough to severely damage a plane's fuselage, and its lightning can take out the electrics. The supercooled water droplets that form in the cloud's upper reaches can coat a plane's wings with ice, fatally altering its aerodynamics and – most dangerous of all – the enormous, turbulent air currents within its central tower can flip an aircraft over like a pancake.

No wonder pilots do all they can to avoid flying too close to these storm clouds. If they can't pass around one, and their plane is capable of flying at high altitudes, they will generally climb over the top. And that is exactly what Lieutenant-Colonel William Rankin, a pilot in the US Air Force, was attempting to do in the summer of 1959 when his jet fighter's engine seized completely and he had to eject. He became the only man to fall through the heart of the King of Clouds and live to tell the horrific tale. His experience made him something of an international celebrity.

Rankin was on a 70-minute routine navigational flight from the South Weymouth Naval Air Station in Massachusetts to his squadron's headquarters in Beaufort, North Carolina. Before take-off, he'd had a word with the meteorologist at the air base, who'd told him to expect isolated thunderstorms *en route*. The thunder-clouds could be expected to reach altitudes of 30,000 to 40,000ft. For a decorated Second World War and Korean War vet like Rankin, this was fairly routine stuff. He knew his jet could reach 50,000ft comfortably and so he was confident of being able to fly over any storms without difficulty. That, of course, was assuming the engine wouldn't conk out just as he was above one.

Forty minutes into the flight, as he was approaching Norfolk, Virginia, Rankin spotted the distinctive shape of a Cumulonimbus ahead. A storm was raging in the town below and the cloud rose in an enormous tower of puffy convection mounds, mushrooming out into a broad, wispy canopy at its top. The summit was at around 45,000ft – somewhat higher than he'd been led to expect by the official back at South Weymouth – so the pilot began a climb to 48,000ft to be sure of clearing it.

Lt.-Col. William Rankin –
before he got intimate with
a Cumulonimbus.

Rankin was directly over the top, at an altitude of 47,000ft and a speed of mach 0.82, when he heard a loud bump and rumble from the engine behind him. He watched in disbelief as the rpm indicator on his dashboard spiralled to zero in a matter of seconds and the bright red 'FIRE' light began flashing urgently.

Sudden and unexplained engine seizure like this is a one-in-a-million kind of emergency and Rankin knew that he would have to act fast. Without power, the jet's controls became ineffective and he instinctively reached for the lever that deployed the auxiliary power package to restore emergency electricity. As he pulled the lever, however, he was horrified to feel it come away in his hands. This sounds like a moment worthy of Buster Keaton, but Rankin was finding it anything but funny. He was wearing just a lightweight summer flying suit. It was unheard-of to eject at this altitude at the best of times. To do so without a pressure suit would surely be suicide.

'The temperature outside was close to –50°C,' Rankin later recounted. 'Perhaps I would survive frostbite without permanent injury, but what about "explosive" decompression at almost ten miles up? And what about that thunderstorm directly below me? If it could be hazardous for an aeroplane in flight, what would it do to a mere human?'[1]

There was little time to ponder the dangers. In a matter of seconds, Rankin realised he had no option but to reach behind his head and yank with all his might on the ejection seat handles. At almost exactly 6pm, he exploded out of the cockpit and began his descent towards the cloud below.

☁

IT IS ESTIMATED that some forty thousand thunderstorms occur around the world each day. At the heart of every one is a Cumulonimbus cloud – often many of them. The cloud can be thought of as Cumulus with ambitions to take over the world. It is what results when a humble convection cloud grows vertically through the mediocris and congestus stages and refuses to stop. A Cumulonimbus can develop from other cloud types too, but does so most commonly from a power-crazed Cumulus like this.

A Cumulus, but turned up to eleven

The classic shape of a mature Cumulonimbus is a huge vertical column, several miles across and extending up as high as 60,000ft (over 11 miles), which spreads out at the top to resemble a blacksmith's anvil. This upper canopy is called the 'incus' (after the Latin for anvil) and consists of ice crystals, rather than the water droplets that make up the rest of the thundercloud. The anvil can stretch out over hundreds of miles, as it is spread by the strong winds high in the atmosphere. From a distance it can have a calm, majestic beauty.

Nick Lightbody (member 95)

You can't miss a Cumulonimbus incus, with its distinctive anvil-shaped top, called an 'incus'.

Barclay Fisher (member 1664)

From afar, Cumulonimbus look calm but, down below, they are anything but.

The same cannot be said for the region below the cloud's central column. With severe winds, hail, lightning and even tornadoes, here a mature Cumulonimbus is a boiling mass of spitting fury. This is where the cloud roars to the world with all the might that the atmosphere has to muster.

Cloudspotters can distinguish a Cumulonimbus from its younger brother, the Cumulus congestus, by careful observation of its upper reaches. If the top of the cloud still has the sharp cauliflower mounds found on a fair-weather Cumulus, it is officially known as a Cumulus congestus. It only becomes a Cumulonimbus when the upper region becomes 'glaciated', which means that its water droplets have begun to freeze into solid ice particles. A Cumulonimbus anvil of ice crystals is brighter and has softer edges than the top of a towering Cumulus.

When clouds are at this scale, it is only possible to judge their overall shape and the appearance of their summits when viewing from some way away. The best distance is around fifty miles. From anywhere near the Cumulonimbus's base, all a cloudspotter will see is a dark, angry overcast sky, from which a lot of precipitation is falling. It is also possible, from this perspective, to confuse the cloud with the dark, precipitating layer cloud called a Nimbostratus. Whilst the Nimbostratus doesn't have anything like the height of a Cumulonimbus, and often spreads out horizontally over hundreds of square miles, it can be hard to distinguish the two from underneath. The weather below a Cumulonimbus is what will give it away. If there is hail, thunder, lightning and strong, gusty winds, then cloudspotters can be confident that they are in the company of the King of Clouds.

From inception to dissipation, an individual Cumulonimbus might last up to an hour or so, leading to a relatively short-lived storm. But thunderstorms can often last much longer than this, since these villains of the cloud world do not always work alone. They have a tendency to form into gangs, which is when they are at their most destructive. As one Cumulonimbus is dissipating, another rises ahead of it. Collectively, they resonate in an enormous self-propagating system of extreme weather that lays waste to whatever is in its path.

The cloud's complex, evolving structure makes it feel almost like a living organism. Indeed, meteorologists describe storms in terms of 'cells' of Cumulonimbus elements. The short-lived storm of an individual cloud is described as 'single cell'. More common, especially in tropical and subtropical regions, are 'multicell' storms, where the convective growth and decay of one cloud triggers the formation of another. Here the series of Cumulonimbus elements combine together in an ordered structure, extending the duration of the storm for many hours.

On occasions, the individual clouds can organise themselves in such a way that they can only really be thought of as an enormous single structure – known as a 'supercell'. These generally form over warm seas, and the updraughts and downdraughts of the Cumulonimbus elements combine into a ferocious self-perpetuating single weather system, stretching over *The cloud* hundreds of miles and lasting for many hours or even a day *beasts are* or so. Supercells are the most common source of large, *alive!* destructive hail, extreme gusts and the devastating winds of the tornado. They are when Cumulonimbus clouds do not act as individual ruffians but, embedded in enormous hurricane systems, exhibit the mob mentality of full-scale riots.

With such destructive capacity, it may seem surprising that the Cumulonimbus is the very cloud that led to the cheery phrase 'to be on cloud nine'. To find out why, one needs to look back to 1896. This is something I am happy to do, since it was of course the 'International Year of the Clouds'.

The year was so named by an international group of meteorologists, brought together by Professor H. Hildebrand Hildebrandsson of the University Observatory of Uppsala in *The Cloud* Sweden and the Hon. Ralph Abercromby of the Royal *Committee* Meteorological Society in London. This pair recruited heavyweights from the meteorological community and gave them the title of the 'Cloud Committee'. Their shared purpose was the business of cloud classification.

The basis for a nomenclature of clouds had been established almost a century earlier by an amateur meteorologist and Quaker called Luke Howard.[2] In 1802 Howard gave a lecture to his local scientific society in which he proposed a classification system similar to the Latin 'Linnean' system based on genera and species, already established in the fields of botany and zoology. Perhaps surprisingly, no one had made any concerted efforts to name specific cloud types before Howard. It was he who coined the names 'Cumulus', 'Stratus', 'Cirrus' and the now defunct 'Nimbus'.

Whilst Howard's system had received swift and far-reaching acclaim, Hildebrandsson and Abercromby were aware that a worrying lack of consistency had developed with the addition of new cloud classifications by various meteorological institutions around the world. They realised that an understanding of the weather depended on co-ordinated observations across national boundaries – something that relied on agreed terminology. With the help of their Cloud Committee and the fanfare of the International Year of the Clouds, they published a pictorial reference book to coincide with the 1896 International Meteorological Conference in Paris.

Published in three languages, the book was called *The International Cloud Atlas* and contained numerous photographs to illustrate the ten cloud genera agreed by the committee. Number nine in the list was Cumulonimbus, the tallest of all the types. To be on cloud nine is therefore to be on the highest one.

Since 1896, there have been seven further English-language editions of *The International Cloud Atlas*, the most recent being in 1995. Now published by the World Meteorological Organisation, it is the undisputed authority on cloud classification, and a book that any serious cloudspotter should own. Sadly, the order of the genera was rearranged in the second edition and Cumulonimbus relegated to number ten. But the 'cloud nine' phrase seemed to stick. It might have been brought into today's popular usage by the 1950s hit American radio show, *Johnny Dollar*, about a freelance insurance investigator. Every time the detective was knocked unconscious, he was transported to cloud nine before coming round again. Later on, of course, in the summer of 1969, the song 'Cloud Nine' was a hit on both sides of the Atlantic for the king of soul groups, the Temptations.

Fig. 23.

As No. 9 in the list of cloud genera published in *The International Cloud Atlas* of 1896, the Cumulonimbus led to the phrase 'to be on cloud nine'.

Besides encouraging agreement on the cloud genera, *The International Cloud Atlas* also established the convention of grouping clouds according to their altitude. The lower region of the atmosphere, the 'troposphere', in which the majority of clouds form, is divided into three levels: low, middle and high. These are sometimes called *étages*, for it was a Frenchman, Jean-Baptiste Lamarck, who proposed a classification system in competition with Howard's that was based on them. The height of the troposphere varies with latitude, but in the mid-latitude temperate regions of the world, the three cloud levels are defined as follows:

LOW-LEVEL CLOUDS form most frequently below 6,500ft.
MID-LEVEL CLOUDS form most frequently between 6,500 and 23,000ft.
HIGH-LEVEL CLOUDS form most frequently between 16,500 and 45,000ft.

Of course the Cumulonimbus King of Clouds refuses to be bound by petty conventions and generally extends through all three *étages*. It feels best to group it with the low clouds, however, since it grows upwards from a base that is always within this level.

☁

'AT FIRST THERE WAS no sensation of falling, only of zooming through the air,' said William Rankin of the moments after he ejected from his stricken jet. Within seconds, he was suffering from the effects of the inhospitable environment at 47,000ft.

'I felt as though I were a chunk of beef being tossed into a cavernous deep freeze,' he remembered. 'Almost instantly all exposed parts of my body – around the face, neck, wrists, hands and ankles – began to sting from the cold.' Even more uncomfortable was the decompression caused by the low pressure at the top of the troposphere as he began the free fall until his parachute would automatically open. He was bleeding from his eyes, ears, nose and mouth as a result of the expansion of his insides, and his body became distended. 'Once I caught a horrified glimpse of my stomach, swollen as though I were in well-advanced pregnancy. I had never known such savage pain.' The one benefit of the extreme cold was that it began to numb his body.

Not the best of flights

In spite of the spinning, flailing nature of his free fall, Rankin managed to secure the emergency oxygen supply to his mouth. It was essential to remain conscious if he was to have any chance of surviving the descent. He was within the upper reaches of the storm cloud, with deteriorating visibility, when he saw on his watch that five minutes had passed since he had ejected. He should have passed the 10,000ft point by now – the height at which the barometric trigger in his parachute would cause it automatically to open. But there was no sign of the parachute. The poor pilot had already suffered an engine failure at 47,000ft, the jet's auxiliary power lever coming off in his hand and having to eject directly over an enormous storm. Now it was beginning to look like he was hurtling through the air strapped to a parachute that didn't work.

Deep within the ice-particle upper region of the Cumulo-nimbus it was dark with zero visibility. This made Rankin totally disorientated, with no idea of his altitude. For all he knew, without his parachute opening he might hit the ground at any moment. It was therefore with great relief that he felt the violent jolt as his parachute finally deployed.

The tension in the risers was enough to reassure him that it had fully opened. He was also relieved to find that, though his emergency oxygen supply had run out, the air at this level was now dense enough for him to be able to breathe without it. In the gloom of the enormous cloud, things appeared to be looking up: 'Under the circumstances, overjoyed to be alive and going down safely, consciously, even the increasing turbulence of the air meant nothing. It was all over now, I thought, the ordeal had ended.' But the turbulence he was beginning to feel and the freezing hailstones starting to strike him meant that he was only now reaching the heart of the storm.

Ten minutes into his descent, Rankin should have been reaching the ground, but the enormous draughts of air that surged up the core of the cloud were retarding his fall. Soon the turbulence became much more severe. He had no visual point of reference in the gloomy depths but he sensed that, rather than falling, he was being shot upwards with successive violent gusts of rising air – blasts that were becoming increasingly violent. And then for the first time he felt the full force of the cloud.

'It came with incredible suddenness – and fury. It hit me like a tidal wave of air, a massive blast, fired at me with the savagery of a cannon... I went soaring up and up and up as though there would be no end to its force.' Rankin wasn't the only one being hurled up and down. In the darkness around him, hundreds of thousands of hailstones were suffering the same fate. One minute they were falling downwards, dragging air down with them; the next minute, they were swept back up by the enormous convection currents within the cloud.

With this falling and rising, the hailstones picked up freezing water and grew in size, hardening layer by layer like gobstoppers. These rocks of ice pelted Rankin with bruising force. He was now vomiting from the violent spinning and pounding and he shut his eyes, unable to watch the nightmare unfolding. At one point, however, he did open them to find himself looking down a long black tunnel burrowing through the centre of the cloud. 'This was nature's bedlam,' he said, 'an ugly black cage of screaming, violent, fanatical lunatics... beating me with big flat sticks, roaring at me, screeching, trying to crush me or rip me with their hands.' Then the lightning and thunder began.

The lightning appeared as huge, blue blades, several feet thick, which felt as though they were slicing him in two. The booming claps of thunder, caused by the explosive expansion of the air as the enormous electrical charge passed through, were so over-powering up close that they were more like physical impacts than noises. 'I didn't *hear* the thunder,' he said, 'I *felt* it.' Sometimes he had to hold his breath to avoid drowning from the dense torrents of freezing rain. At one point he looked up just as a bolt of lightning passed behind his parachute. It lit up the canvas, which appeared to the exhausted pilot as an enormous, white-domed cathedral. As the image lingered above him, he thought that he had finally died.

How it feels to be a hailstone

☁

THREE CRITICAL CONDITIONS of the atmosphere provide an ideal environment for a fledgling Cumulonimbus cloud to grow into a large, angry specimen:

1) There needs to be a ready supply of warm, moist air around the cloud. This acts as the energy source, fuelling the cloud's growth. The central core contains enormous updraughts of air, rising at speeds of 25–70mph, and the stream of air that supplies this updraught is known as the 'inflow'. When the inflow air is warm and moist, plenty of heat is released as it forms water droplets within the cloud. This energy gives the air at the centre buoyancy, increasing the updraughts and the cloud's growth.

2) The tropospheric winds around the Cumulonimbus need to increase considerably with height in the direction of the cloud's movement so as to encourage it to slant forward. This is critical for the cloud's longevity because its central tower is not just the region of violent updraughts; it is also the part where the heavy precipitation, such as hail, develops. As the precipitation falls through the cloud it can chill the air by partly evaporating and in addition drags the air down with it. The plunging downdraught so formed can, if the cloud is vertical, swamp the life-giving

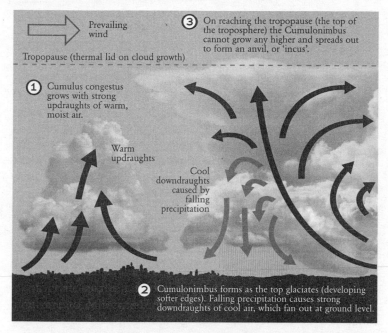

The top of the troposphere, which acts as an invisible lid on the growth of clouds, is what causes the Cumulonimbus to spread out in an anvil shape.

updraught to kill off the cloud quite quickly. These downdraughts reach the surface to spread out like water poured on a table – often causing an advancing line of low cloud at the leading edge. However, when the surrounding winds make the Cumulonimbus slant forward, the precipitation falls slightly ahead of the updraught, reducing its tendency to cancel out the rising core and thereby kill the cloud's growth.

3) The atmosphere around the cloud needs to be 'unstable'. This has to do with the degree to which the air becomes colder with altitude. If the temperature of the surrounding environment decreases steeply with height, then the warm, moist air entering at the inflow and cooling as it rises will always be a little warmer than the air around and so will remain buoyant. This is what encourages the growth of the cloud. The troposphere always tends to become cooler with altitude, but it often does so more dramatically near the

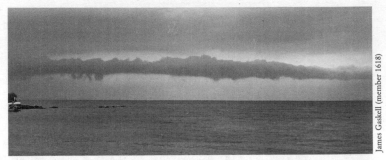

A 'roll cloud' can sometimes appear in advance of a storm. It forms as the cool downdraughts spread out at ground level and push up the warmer air ahead.

surface in tropical regions where the warmer ground increases low-level air temperatures. This is one reason why thunderclouds are so common in these parts of the world.

Incidentally, it is the way that the air temperature changes with altitude that gives the Cumulonimbus its distinctive anvil shape. The top of the troposphere is defined as the part of the atmosphere where the air always stops cooling with height. Known as the 'tropopause', it is a layer where the temperature remains constant – say, at around –50°C – before starting to increase again in the lower stratosphere. This change in the temperature gradient acts as a thermal ceiling to cloud growth. A Cumulonimbus reaching it is unable to grow any taller and so spreads out under the ceiling.

Just as Cumulus clouds can occur in the different species of humilis, mediocris, congestus and fractus, the Cumulonimbus can be one of two possible species. These are called 'calvus' and 'capillatus' and they are distinguished by the appearance of the upper, ice-particle region. Cumulonimbus calvus is when the cloud's anvil is smooth with soft edges. Cumulonimbus capillatus is characterised by an upper region that is fibrous and striated. It is named after the Latin for hair, and can look like the disorderly locks of a child who's just been in a playground scrap.

It should come as no surprise that the King of Clouds prefers not to travel alone. Besides the incus anvil at its top, the Cumulonimbus has a whole court of 'accessory clouds' (clouds that only ever appear near or merged with one of the ten genera) and 'supplementary features' (various forms and protuberances attached to one of the cloud genera). These act like an entourage.

The wide, trunk-like 'wall cloud' is one that forms below the Cumulonimbus base, around the core of its updraught region. 'Pannus' are dark, ragged shreds of cloud that appear below the storm cloud's base, as the air becomes saturated from the heavy precipitation. Ahead of the storm, riding on the front of the outflow of cool air, there can appear a dense shelf or roll of advancing cloud, called an 'arcus'. The 'pileus' cloud is one that can appear as a smooth veil or cap over the Cumulonimbus's summit. It forms when a high layer of moist air is forced upwards by the rising central tower and rarely lasts long before the Cumulonimbus grows through it and it becomes subsumed into the body of the main cloud. The 'velum' forms in a similar way, but is a large, flat patch of soft-looking cloud, which forms when a series of towers from separate clouds act together to push a large region of moist air upwards. The velum can hang around for some time after the Cumulonimbus clouds have dissipated away. Then there is the 'tuba', which is the first sign of a tornado developing below the Cumulonimbus. It is a finger of cloud that lowers from the cloud base, forming in the centre of a vortex, and results from the cooling of the air in the lowered pressure of the spin.

Most dramatic of all the court attendants are the breast-like 'mamma' (which are sometimes called 'mammatus'). These are udders of cloud that can hang on the underside of the Cumulonimbus's anvil and indicate high instability in the air around the top of the cloud. They are associated with particularly violent storms. Finally, there is often a line of growing Cumulus congestus, queuing up along the storm cloud's inflow. These are the pretenders to the throne – ready to step in and assume rule the moment the king expires.

Amidst the mêlée of its fussing court, the Cumulonimbus itself is possessed by an all-consuming fury, which is fuelled by the unstable atmosphere of its reign. How fitting that the events of Shakespeare's *King Lear* should unfold against the backdrop of a raging tempest, for Lear was driven mad by his own unstable atmosphere.

Graeme Ferris (member 1591)

The mother of all clouds – 'mamma' are the breast-like features that can appear
on the underside of the Cumulonimbus's anvil.

LEAR: *Blow, winds, and crack your cheeks! rage! blow!*
You cataracts and hurricanoes, spout
Till you have drench'd our steeples, drown'd the cocks!
You sulphurous and thought-executing fires,
Vaunt-couriers to oak-cleaving thunderbolts,
Singe my white head! And thou, all-shaking thunder,
Strike flat the thick rotundity o' the world!
Crack nature's moulds, all germens spill at once
That make ingrateful man![3]

Of course, Lear's insanity was due to estrangement from his
daughters and alienation from the kingdom he was relinquishing
rather than the unstable temperature profile of the troposphere.
That aside, he does a pretty good impression of a Cumulonimbus.

☁

THE ANCIENT POETS OF INDIA always saw the start of the
monsoon season as a time of great romance. They regarded it much
as the romantic poets of Europe did the spring. Monsoon rains
bring relief from the scorching heat of India's summer, cause the

gardens to spring to life with abundant colours and fragrances, and inspire the wild peacocks to begin their flamboyant courting rituals. The messengers that herald this transformation are none other than the Cumulonimbus clouds, which has earned them a special place in Indian hearts. Nowhere has this been more beautifully expressed than in a poem by the greatest Sanskrit poet, Kalidasa, written sometime between 50BC and AD400.

It is called *The Meghaduta*, which means 'The Cloud Messenger', and is about a 'Yaksha' – one of the demi-gods – who was responsible for guarding the treasures and gardens belonging to Kubera, the Hindu god of wealth. This unnamed Yaksha had not fulfilled his duties well – perhaps he had forgotten to lock up the god's store of unimaginably valuable jewels, this isn't specified – and so his master put a curse on him, banishing him from his home in the Himalayas to spend a year alone in the Vindhya Mountains of central India.

Clouds performing errands

Wandering aimlessly from one mountain hermitage to the next, the Yaksha had nothing to do but pine for his wife back home and count the lonely months till he could return. Eight months into his exile, he noticed a dark Cumulonimbus thundercloud clinging to the peak of the mountain. It could mean only one thing.

The Yaksha knew that the monsoon season was when travelling men returned to their wives, and the sight of the cloud made his longing all the more acute. Noticing that the southerly wind would blow the cloud in the direction of his home in the Himalayas, he decided to ask it to carry a message to his wife.

> *You are the refuge, O cloud, of the afflicted: Bear, therefore, to my beloved a message from me, separated from her by the anger of the Lord of Wealth...*
> *When you appear in the sky fully equipped, what other man, whose life is not dependent upon another like mine, will neglect his wife afflicted with separation?* [4]

He gave the storm cloud detailed directions to his home city in the north. He pointed out rivers along the way where the cloud could stop and drink, and suggested mountain peaks that it could embrace for rest. The Yaksha gave lyrical descriptions of the scenes the cloud would encounter on its journey. In the town of Ujjayini, for instance, it would notice girls dancing at the shrine of Siva. Feeling 'the first drops of rain-water shooting to their nail-marks', they would glance up excitedly, for they knew that their own lovers would soon be coming home.

The Yaksha explained how to find his house, upon reaching his home-town. His wife would be sitting inside, unable to sleep or eat. No doubt, she would be so consumed by longing for her husband that she would be looking a bit rough:

> Surely the face of that my beloved, resting on her hand, having its eyes swollen by excessive weeping, with its lower lip changed in colour on account of the warmth of her sighs and partially visible owing to her hair hanging loosely about it, bears the miserable appearance of the moon whose light is obscured by your interruption.[5]

He warned the thundercloud not to startle her, but to rouse her with a breeze, cooled by its droplets, and to be sure that its lightning was suppressed in its interior. He asked it to console her with low thunder, to tell her that she shouldn't give up waiting for him, for the curse that kept him away would soon cease. Having delivered its message, the cloud would be free to drift off and enjoy the splendour of the rainy season.

☁

KALIDASA'S YAKSHA leaves his Cumulonimbus messenger with one parting wish: that the thundercloud will never have to suffer cruel separation from its own beloved consort, the lightning. Who can know the workings of another's marriage? Who can say what it is that keeps the spark alive or, indeed, why a seemingly innocuous comment can start an explosive row? This applies as much to the marriage between a Cumulonimbus and its lightning as it does to any other. Lt.-Col. William Rankin would have been the first to attest that the inside of a thundercloud is not an environment conducive to sober

The intimate secrets of a Cumulonimbus's love life

Ashley Gibbs (member 563)

When Cumulonimbus clouds show off.

observation and measurement. Amongst all the confusion, it is very hard to predict when and where a bolt will strike, making lightning particularly difficult for scientific study.

Thunder, by comparison, is now well understood. It is not, as Aristotle maintained, the sound of a 'dry exhalation' ejected from a cooling cloud stack hitting the surrounding clouds. Nor is it, as René Descartes suggested, caused by the air between two clouds resonating like that in an organ pipe, as one descends on to the other. I do, however, like to imagine that there is a raging debate amongst meteorologists as to whether it is Raiden, the Japanese god of thunder and lightning, who creates the sound. He looks like a red demon, has sharp claws and is fond of eating human navels. Not wanting to take any chances, young children in Japan put their hands over their bellybuttons when they hear thunder.

Alas, that raging debate does not exist, however – we know that the extreme heat of the lightning bolt is what produces the crashing rumble of thunder. The bolt instantaneously heats the air to 27,700°C. This is more than four times as hot as the surface of the Sun, and the heating takes place within a few millionths of a second, causing an explosive expansion of the air around the channel of the bolt. The resulting waves are what we hear as the tearing clap of thunder.

Our understanding of the finer points of lightning formation is patchy. But the basic principles are clear. It is electricity passing through the air, due to a cloud developing regions of differing electrical charge.

Charge can build up in a Cumulonimbus cloud in a way that is comparable to that which you pick up while walking on synthetic carpets. Your shoes collect electrons from the fibres of the carpet, causing a charge imbalance between you and your surroundings. When you touch a conductor, such as a doorknob, the negative charge flows away from your finger in a spark.

A Cumulonimbus cloud doesn't wear shoes, but it can develop a charge imbalance from collisions between the large hailstones and smaller ice particles within the highly turbulent environment of the storm cloud. It so happens that as they collide, the large hailstones tend to pick up electrons (which are negatively charged) from the smaller ice particles. By doing so, they gather negative charge, while the smaller particles are left with a corresponding positive one.

The cloud's convection updraughts cause the smaller particles to be wafted towards the top, while the heavier hailstones fall towards the bottom. In this way, the Cumulonimbus can develop a charge imbalance without going anywhere near a synthetic carpet.

The separation of charges within the cloud is hardly a stable state of affairs. The violent convection currents in the heart of the Cumulonimbus have led to a state of marital tension and things are all set for a flare-up – one that takes the form of an enormous redistribution of electricity.

☁

WHEN I WAS YOUNG, my family lived in a block of flats with a view over the roofs of West London. Tower blocks are always great vantage-points for cloud gazing, and I made full use of this whenever there was a thunderstorm after bedtime. Standing in the dark between the curtain and the cold pane of my bedroom window, I'd peer through the downpour towards the centre of the storm and try to work out where, amongst the swirling mess of rain and cloud, the next bolt of lightning would strike. Wherever I directed my gaze, it always seemed to flash somewhere else.

Whilst waiting for the next stroke, I would draw shapes in the condensation of my breath on the windowpane and watch the rivulets of raindrops join and divide as they ran down the outside. I'd try to predict whether the next flash would be a forking streak or just a general illumination of the cloud – whether it would be 'fork' or 'sheet' lightning. In actual fact, there is no difference between the two – sheet lightning is merely when the body of the cloud hides the fork lightning from view, and one sees a flickering illumination of the cloud as a whole.

There are, however, some very real ways that one bolt of lightning can differ from another – the most fundamental being where the bolt goes to. Though we typically think of lightning passing from 'cloud to ground', this is only one of the possible routes. Indeed, cloud-to-ground isn't even the most common type of lightning.

That is known as 'in-cloud' lightning – where the spark travels from one part of the thundercloud to another, thereby evening out the imbalance of electrical charges within the cloud. Less common than both types is 'cloud-to-cloud' lightning, where the discharge is from the negative region of one cloud to the positive one of its neighbour. The type that is least obvious to an observer on the ground, and is certainly least understood, is 'cloud-to-air' lightning, where the charge travels between the top of the cloud and the atmosphere above.

High-speed footage of lightning bolts striking the ground reveals that the charge moves in several distinct stages. A jagged, forking 'stepped leader' of negative electric charge descends from the cloud, but before one of its spikes reaches the ground, an 'upward streamer' of positive charge emerges from below to meet it. When the two connect, a circuit is formed and the channel lights from the bottom up as the electricity rushes down to redistribute the charge. This is the flash itself, and is called the 'return stroke'. Lightning appears to flicker when charges from other regions of the cloud pass down the same channel right after the first, causing successive return strokes.

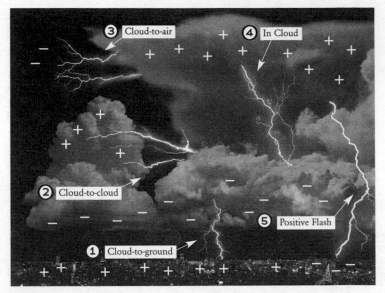

This schematic diagram shows possible routes that lightning bolts can take between regions of negative and positive charge.

Whilst these are the basics, nothing that involves the Cumulonimbus cloud is ever straightforward. What, for instance, triggers the lightning to strike at one particular moment and place, rather than another? What are the spheres of light, known as 'ball lightning', which observers report as being the size of a grapefruit and drifting around at ground level for a matter of seconds during thunderstorms? In the words of Martin Uman, who leads the University of Florida's Camp Blanding International Center for Lightning Research and Testing, and is a world expert on the subject: 'There is an awful lot we still don't know about lightning.'

☁

WITHIN THE LAST couple of decades, it has emerged that a family of mysterious electrical phenomena sometimes form in the atmosphere way above very large storm systems. Like so many advances in science, their discovery was a complete accident. In 1989, John R. Winckler, a professor at the University of Minnesota, was testing a very sensitive low-light video camera for a rocket launch. On reviewing the tape, Winckler noticed a frame that appeared to capture a giant column of light rising above a thunderstorm near the US–Canadian border. He showed it to a colleague, Walt Lyons, who was developing a lightning detection network at the university, and the two decided that this was no technical fault. It appeared to be some hitherto unidentified electrical discharge.

In the following years, Lyons became a world authority on the tricky task of capturing these electrical phenomena on film from an observation platform at his home on the Great Plains of *Sprites, elves* Colorado. For many years, scientists couldn't agree what to *and blue jets* call them. It wasn't until 1994 that one professor's use of the name 'sprite' became accepted, as it seemed appropriate for beautiful, fleeting, magical phenomena about which we understood so little. They appear for only one tenth of a second – just long enough to be seen with the naked eye – and are often in the shape of gigantic red jellyfish, with bluish tinges in their descending tendrils.

Starting at an altitude of around 45 miles, sprites rise to around 55–60 miles and descend to some 15–20 miles. Photographs suggest that they don't even touch the clouds below. They occur primarily above giant storm systems and appear immediately after bolts of a particular type of lightning below. Called 'positive cloud-to-ground lightning', these are not typical lightning strikes, accounting for only 5–10 per cent of all strikes. While jellyfish are most common, sprites form in a whole range of joyful shapes, which have earned them names like 'broccoli sprites', 'octopus sprites' and 'Carmen Miranda sprites'.

To this day, scientists still don't agree on what exactly they are, especially since they occur in the region above the troposphere and stratosphere,

A 'Carmen Miranda sprite'. Sprites are mysterious electrical displays that can appear 50 miles above a thunderstorm.

called the 'mesosphere' – one that was always thought to be electrically inert.

Following Winckler's chance sighting, teams of atmospheric researchers have made pilgrimages to Lyons's observation centre, which has become known as 'Sprite Central'. With ground-based cameras, as well as those on aeroplanes and the Space Shuttle, researchers have identified two other apparently related forms of electrical discharge above thunderstorms. These have been given equally evocative names: 'elves' and 'blue jets'.

Elves are not visible to the naked eye as they last less than a thousandth of a second but, like sprites, they appear at the same time as positive cloud-to-ground lightning strikes. If they were not so short-lived as to be invisible, they would probably also appear red. Taking the shape of gigantic expanding doughnuts, elves form 60–65 miles up in the atmosphere and extend outwards to several hundred miles in diameter.

Blue jets are just visible, and spurt upwards from the top of the Cumulonimbus cloud at speeds of 50–100 miles per second, reaching heights of up to 25 miles before fading. They, like elves, are much less common than sprites. Whilst they do not appear to be related to specific cloud-to-ground lightning bolts, it seems they are generated above storms with high lightning rates.

As speculation amongst the scientific community continues over how these mysterious electrical offspring form at such heights, one thing is certain. There is a great deal more to the Cumulonimbus's relationship with its lightning spouse than meets the eye.

☁

CLOUDSPOTTERS WILL BE PLEASED to learn that Lt.-Col. William Rankin didn't die. After his vision of the parachute billowing as a cathedral above him, he began to notice the air becoming less turbulent, the rain and hail losing intensity. He was finally emerging from the underside of the cloud.

In spite of his ordeal, Rankin managed to land successfully in a forest of pine trees. The storm was still raging, but on the ground it was nothing compared with what he'd experienced above.

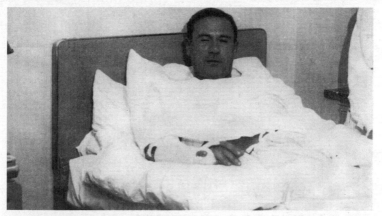

Rankin recovering in hospital. He took cloudspotting to new heights.

Finding that his limbs were not broken, the pilot managed to pick himself up and stumble in search of a road for help.

When he was later examined in the hospital at Ahoskie, North Carolina, the doctors reported that his body was discoloured from frostbite and covered in welts and bruises from the impact of the hailstones. His torso also showed impressions of the stitching of his flight jacket, which had strained against the expansion of his insides in the severe decompression after ejecting. The doctors were as amazed as Rankin was that he'd survived.

Moments after the pilot had landed in the forest, he had peered through the dim light of the storm and just made out the fluorescent hands on his watch. Under normal conditions, a parachute descent from 47,000ft could be expected to take around ten minutes. Given that he had ejected from his jet at exactly 6pm, he was stunned to see the watch read 6.40pm. Rankin had been buffeted up and down by the violent turbulence of the Cumulonimbus for a full forty minutes – no more than a pilot-shaped hailstone in the icy heart of the King of Clouds.

THREE

STRATUS

The low, misty blankets

Stratus is a flat, grey, indistinct sheet of cloud, which is generally featureless in appearance. There are no crisp-looking cauliflower mounds to catch the sunlight on this formation – just an overcast veil that casts a drab, dreary light.

Compared with the frenetic and capricious convection clouds, the Stratus is a ponderous individual. It rarely bothers to shed much of its moisture – never managing more than a light drizzle or a gentle snow. It takes its time arriving, and generally outstays its welcome when it does. This is not a cloud known for its spontaneity – it isn't the type to cause a commotion at picnics with a sudden downpour the moment the sandwiches are out of their foil. When there is a thick layer of Stratus above, people are just more likely to forget the picnic and opt for the cinema instead.

As founder of The Cloud Appreciation Society, I like to think that my affections extend to each and every cloud formation. On a chilly February morning in London, however, there is a Stratus cloud above that is weighing heavily on my mood.

Many clouds give a sense of depth and scale to the sky – one that is lacking on a clear day. Looking up at the different layers and formations, a cloudspotter is able to appreciate the expanse of the heavens. But this Stratus is making me feel claustrophobic. And that is not good when you are outdoors. It is reminiscent of an acquaintance who insists on standing too close and invading your

HOW TO SPOT
STRATUS CLOUDS

Stratus are grey layers or patches of cloud, with very diffuse edges. They are the lowest-forming of all the cloud genera, sometimes appearing at ground level, when they are called fog or mist.

TYPICAL ALTITUDES*: 0–6,500ft
WHERE THEY FORM:
Worldwide. Most commonly around coasts and mountains.
PRECIPITATION (REACHING GROUND):
No more than occasional drizzle, snow or snow grains.

Stratus nebulosus translucidus

Stratus fractus

STRATUS SPECIES:

NEBULOSUS: By far the most common, when it is in a grey, generally featureless layer.

FRACTUS: When it is in separate, ragged shreds of grey cloud. This can appear in the region below precipitating clouds, when it is called 'pannus'. Though not particularly thick, these shreds can look quite dark against the base of the cloud above.

STRATUS VARIETIES:

OPACUS: When the layer is thick enough to completely mask the Sun or Moon.

TRANSLUCIDUS: When it is thin enough to show the outline of the Sun or Moon.

UNDULATUS: A rare variety, in which the layer has wave-like undulations to its surface. The surface of Stratus is rarely distinct enough for this to be observed.

NOT TO BE CONFUSED WITH...

CIRROSTRATUS: which is a high layer cloud that can look similar to a very thin Stratus. Being made of ice, however, it has a whiter tone.

ALTOSTRATUS: which is a mid-level layer cloud, often consisting of droplets, like Stratus. Through a layer of Stratus, the outline of the Sun (when it is discernible) is less diffuse, compared with the 'ground-glass' appearance of the Altostratus.

NIMBOSTRATUS: which is a thick, dark layer of precipitating cloud that might be confused with a thick Stratus. But this has a less ragged base than the Nimbostratus and produces lighter precipitation.

* These approximate altitudes (above the surface) are for mid-latitude regions.

personal space. Not only that, the cloud leaves me without the faintest idea of where the Sun is.

I am on my way to the office through the capital's grey streets and, though it is morning, the sky looks no different from how it would in the afternoon. This Stratus is of the variety known as 'opacus', which means it completely hides the Sun. Were it 'translucidus', I would at least be able to see a brighter glow or a soft outline showing its position.

Besides being described as opacus, this Stratus is of a species called 'nebulosus', which means that there are no variations to its tone, no lighter and darker patches, no visible clumps or shapes to its underside, just a formless, featureless concrete grey, stretching as far as the eye can see. Given that the Stratus is a low cloud, that is not particularly far. I'm walking below a sky of stagnant dishwater. No wonder it is making me feel miserable.

This is, of course, potentially damaging news. Imagine the international scandal if it emerges that the head of The Cloud Appreciation Society is down on account of a Stratus cloud. I may be a lover of clouds but today I long to catch a glimpse, just a fleeting glimpse, of brilliant sunshine. Below this ceiling of Stratus nebulosus opacus, it feels as if God's decided to cut his fuel bills and install a fluorescent strip.

☁

CLOUDSPOTTERS SHOULD KNOW that the featureless nebulosus is one of just two possible species of Stratus. The other one is Stratus 'fractus'. This is when the low cloud layer is broken into blurred patches or shreds. Unlike the nebulosus, which just hangs there for long periods, the Stratus fractus has outlines that are constantly changing. It is essentially the same as the pannus accessory cloud that can form in the saturated air below a Cumulonimbus.

Besides being divided into species, each cloud genus can have a number of possible 'varieties', referring to common characteristics that the cloud may exhibit. In addition to opacus and translucidus, which depend on whether you see the position of the Sun or Moon

Stratus nebulosus – it's never known to make you feel elated.

through the layer, Stratus has a third variety called 'undulatus'. This is when the layer's surface has a wave-like appearance, which can result from the action of the winds at cloud level.

Stratus forms in an entirely different way from the Cumulus and Cumulonimbus convection clouds. Like all clouds, it appears when air cools enough for some of the water vapour it holds to condense into liquid droplets. But in this case, a large area cools, rather than individual pockets cooling as they rise in thermals.

How does a layer of air cool like that? One of the ways is for the whole lot of it to rise at the same time. This can happen if the air drifts into contact with a region or area of slightly cooler air. The cooler air, being denser, tends to remain near to the ground and the warmer air can gently rise over it. The layer cools as it rises and drops in pressure and, if the whole process happens in a gentle way, this cooling can result in a smooth, featureless blanket across the heavens. Stratus is associated with relatively stable air, compared with the turbulent currents of the convection clouds. That is why it has a tendency to hang around, as it is doing today on my way to work.

Whilst light rain or snow can sometimes fall from Stratus, this one is not raining at all. I would feel better if it were. A vigorous downpour is an invitation to put the fire on and enjoy that cosy in-here-versus-out-there feeling. But that is not the style of the Stratus nebulosus opacus. When a Stratus appears to be raining hard, this is usually just because it is shielding a higher rain cloud from view.

What makes this cloud stand out from the cloud family is the very fact that it doesn't stand out. I love clouds for the endless variety they bring to our skies. Life would be dull if we had to look up at blue sky day after day. The playful fair-weather Cumulus and the angry, mighty Cumulonimbus are in a constant dance of growth and decay. 'Huge cloudy symbols of a high romance,' as Keats put it; they are Nature's poetry, writ large for all to see. This Stratus nebulosus opacus is, by contrast, profoundly unpoetic.

I remember looking up at the same clouds lying on my back in the middle of a busy road. I was seventeen, and had just been knocked off my motorbike. My mother always said the accident happened because family difficulties at home made me take my eye off the road. As I lay on the Tarmac, my leg twisted at a horribly inappropriate angle, I looked up at the sky and waited for the ambulance to collect me. Above was a thick, overcast Stratus just like the one today: low, grey, oppressive...

Not long after, my friend's father, Neville Hodgkinson, told me something that seemed pertinent. He practises the eastern religion of Raja Yoga, as taught by the Brahama Kumaris, an *Spiritual* organisation established in Hyderabad in 1937. Neville told *distractions* me that clouds have a symbolic role for some yogis: they stand for the times when the yogis lose track of their spiritual journey. They symbolise the distractions that come between the yogis and the 'Supreme Light' of God.

He didn't say what these distractions were, but given that the Brahama Kumari yogis are teetotal vegetarians who avoid garlic (which they consider inflames the carnal passions) and practise a strict vow of celibacy, a few possibilities came to mind. Anyway, these distractions occasionally become so profound and sustained that the yogis lose track of their spiritual path altogether. They call it a 'storm of Maya'. It is one in which illusory ways of thinking and feeling block out the Supreme Light altogether. At times like these, he said, the yogis remind themselves that, beyond the clouds, the Sun never stops shining.

This brought to mind the revelation I had in an aeroplane as a child: that for the pilot it was, without exception, always a sunny day at work. Not only that, but the view from his office window was one of constantly varied and beautiful cloudscapes. So what about the rest of us, stuck down here, terrestrial in our jobs and distracted in our spiritual paths? On a bleak, grey February morning, it can sometimes be hard not to yearn for the limpidity of direct sunlight.

☁

PERHAPS I AM SUFFERING from Seasonal Affective Disorder (SAD)? It is a syndrome that was identified by Dr Norman Rosenthal, a clinical psychiatrist based in North Bethesda, Maryland, in the USA. Dr Rosenthal defines it as a combination of depressed mood and a characteristic cluster of physical symptoms that is recurrent with the changing seasons. 'Winter SAD' sufferers find that, besides feeling down in the darker months, they tend to have less energy than usual, feel less creative and productive, need more sleep and have less control over their appetite.

I have experienced these symptoms, but usually only after a big night out. I can't say that there is anything seasonal about them. Amongst those who can, the number of sufferers has been found to increase with latitude. The further they are from the equator, the more people complain of winter SAD symptoms.

Dr Rosenthal has found that the number of sufferers in the USA ranges from 1.4 per cent of the population in Florida to 9.7 per cent in the higher latitudes of New Hampshire. Women are more prone to it than men, and there is evidence that this discrepancy might be hormonally related, since it increases after puberty and decreases in the post-menopausal years.

Attributing the trigger for winter SAD symptoms to be the amount of light the individual sees, Dr Rosenthal has found that when a sufferer sits in front of a 10,000-lux light box for thirty minutes each morning – perhaps while reading or doing paperwork – there is often a marked improvement in mood and energy.

Well, I'm not about to sit in front of a light box every morning

Is that a factory poking through the Stratus?

just because of a bit of Stratus nebulosus opacus. The problem is not so much the lack of light, but the fact that it can mean there is nothing to look up at all day long.

Could not the same be said of relentlessly blue skies? Dr Rosenthal has also identified a version of Seasonal Affective Disorder that he calls 'summer SAD'. Those who suffer from this actually feel down during the summer months. Interestingly, winter types are far more common in the United States and Europe, while in Japan and China more people have been found who have a summer bias to their seasonal depression.

In Britain, unlucky or unhappy people are said to have a 'cloud hanging over them', whereas those with an optimistic demeanour have a 'sunny outlook'. Office brainstorms, where no one is allowed to criticise stupid ideas, are described as 'blue-sky thinking'.

In Iran, by contrast, you would remark that someone is blessed or lucky by saying 'dayem semakum ghaim,' which translates as 'your sky is always filled with clouds'. To a nation whose skies are clear and blue for months on end, there's nothing special about having a sunny outlook; no great merit to blue-sky thinking. There, the clouds are both a promise of precious rain and a blessed relief from the baking Sun. In temperate regions, where rain is rarely scarce, feelings towards clouds seem more confused. On the one hand, they obscure the life-giving rays but, on the other, they are an endless source of beauty. What, after all, is a sunset without the clouds? A bright ball disappearing behind a line, that's what.

It feels as if there will be no more sunsets at all if this Stratus nebulosus opacus continues to veil the sky. Not only is it the annoying friend who stands too close, it's also the one who doesn't know when it is time to leave.

Wait a minute… Something's just occurred to me about the Stratus. And, I'm glad to say, it's enough to make me more than willing to forgive its oppressive ways.

WITHOUT THE STRATUS, I would never have experienced the peculiar joy of walking through a cloud. Being the lowest of all the types, whose base rarely forms above 1,600ft, it is the only one that happily comes down to join us at ground level. Tethered to terra firma like this, Stratus is referred to as fog or mist.

When I was young, there was nothing more entrancing than waking to find the world outside had been draped with mysterious veils of fog. No storms or sudden freezes heralded the transformation. The fog arrived unannounced – 'on little cat feet,' as the American poet, Carl Sandburg, put it:

> It sits looking
> over harbor and city
> on silent haunches
> and then moves on.[1]

I loved the way the gentle mists changed everything. Out in the garden our pet cat, Pepsi, seemed to emerge gradually from the blur as she walked up the path and yet, at the same time, to take form abruptly. I loved the way sounds were transformed. Disembodied

Stratus is the only cloud that bothers to come and join us down at ground level.

voices seemed both miles away and beside me at the same moment. Without the Stratus cloud, I would never have experienced the veiled magic of foggy mornings.

Victor Hugo wrote, 'Woman, nude, is the blue sky. Clouds and garments are an obstacle to contemplation. Beauty and infinity would be gazed upon unveiled.'[2] It strikes me that clouds and garments are both instruments of seduction. As 'obstacles of contemplation', each can stimulate our appreciation of beauty – in one case, that of the human form, in the other, of the sky. To walk through fog is to experience at first hand the seductive veils of the clouds.

Beauty should be painted with her head lost in the clouds, wrote Cesare Ripa in *Iconologia*, his sixteenth-century guide to iconography in art and sculpture, for there is nothing about which it is more difficult to speak in a mortal language, or that can be less easily understood by human intelligence.

☁

THE SENSATION OF walking through fog inspired the design of the centrepiece pavilion at the Swiss National Expo of 2002.

As visitors approached the building along a 400ft ramp bridge over Lake Neuchâtel, near Yverdon, all they could see was a mass of fog, seemingly settled on the surface of the water. The Blur Building was designed by the New York architects Liz Diller and Ric Scofidio, and was shapeless, dimensionless and surfaceless. It had no walls or roof to speak of. It was a building made out of a cloud. And if we're going to get technical about it, which I think we should, it was constructed from a surface-based Stratus cloud.

Diller and Scofidio won the competition to design the pavilion in 1999. Their proposal was to erect a metal skeleton, suspended above the surface of Lake Neuchâtel, which would be covered in extremely fine precision nozzles. Water would be pumped up from the lake below, filtered, and sprayed into the air to create the building itself. The Blur Building would be a challenge to the whole idea of what a building is, but it wouldn't be the first architectural use of man-made fog.

In 1970, at the Osaka World's Fair, the Japanese sculptor Fujiko Nakaya used water jets to shroud the geodesic dome of the Pepsi Pavilion in cloud. But hers was a building with a solid shell. Diller and Scofidio were proposing to take the idea a step further. The Blur Building would have no hard shell – like a real cloud, coaxed from the heavens to the lake surface, it would expand and contract with the prevailing winds and humidity over the six short months of its existence. It would also prove a nightmare to build.

Nevertheless, in May 2002 the jets were turned on, the cloud took form, and the Blur Building was opened to the public. Its metal skeleton was covered with 31,400 high-pressure water jets. As the water was forced through them, it was 'atomised' into tiny droplets, measuring only 4–10 thousandths of a millimetre across – around the same size as real fog droplets.

The water pressure was controlled by a sophisticated computer system, which took into account the temperature and humidity of the surrounding air as well as the prevailing winds. To ensure that the structure wasn't stripped of its garments on a windy day, the nozzles on the windward side sprayed more water than those on the leeward. In high winds, the fog – or perhaps I should say the building itself – would stretch out beautifully across the lake. By

Diller and Scofidio's Blur Building
at the 2002 Swiss Expo was clad in
a Stratus cloud.

responding dynamically to the constantly changing atmospheric conditions, the system ensured there was always enough fog to envelop the structure, but not so much as to cause a nuisance downwind.

Visitors walked along the bridge over the lake, leaving the reassuring solidity of the shore behind. The other end disappeared into the swirling depths. It was both disconcerting and thrilling to step into the formless blur.

The architects theorised about their project like architects always do: 'To "blur" is to make indistinct, to dim, to shroud, to cloud, to make vague, to obfuscate,' explained Diller. 'For our visually obsessed, high-resolution/high-definition culture, blur is equated with loss… The clouding of vision carries many associations. I think of Jack the Ripper, London, and killers lurking in the fog. The fogged or the veiled is always menacing.' What she was trying to say is that it's exciting to walk through fog.

Construction of the Blur Building had been fraught with

difficulties. Apart from the huge technical challenge of creating and controlling the fog itself, the budget was slashed half-way through the build. At times, it looked as if the project was going to go belly up. Matters weren't helped by ill-prepared and disappointing fog-making tests in front of the assembled world media. 'The flop of the world's most expensive cloud,' howled *Blick*, Switzerland's mass-circulation newspaper. 'For ten million we only get FOG!'

When it finally opened, however, the public greeted the Blur Building with delight, and all the difficulties seemed to be forgotten by the press. 'What a crazy, idiosyncratic thing!' said the *SonntagsZeitung* in May 2002. 'How deliciously without purpose!... The cloud has enchanted the whole country.'

WHEN IS EARTH-BOUND Stratus described as fog, and when is it mist? The official distinction relates to how far you can see through it. If you can see less than one kilometre, then meteorologists call it fog. Visibility between one and two kilometres, and they call it mist. (If you can see less than 1km and there is no Stratus, then you are just short-sighted.)

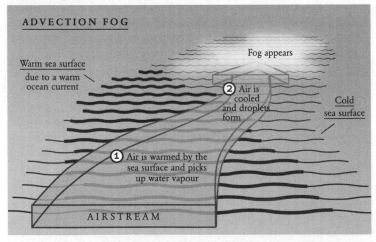

'Advection fog' can form when air passes from a warm sea surface to a cold one.

The two differ in terms of the size and density of their water droplets. In explaining how they form, I'll just talk about fog. They form in the same ways, so you can substitute the word 'mist' when and where you feel the inclination.

There are two main types of fog: 'advection' and 'radiation'.

When Stratus clouds turn bad...

And these certainly do form in different ways. Advection fog is like the one that came rolling in with menacing speed over Antonio Bay in the 1980 horror movie, *The Fog*. This was not a great film, which is a shame, as it is the only Stratus-based horror I've seen. It does, however, provide a helpful introduction to the sort of fog that forms by the process known as advection.

At midnight, on the eve of its hundredth anniversary, the little coastal town was engulfed in a thick fog that swept inland from the sea. What was particularly worrying for the locals was that it contained a 'deadly evil'. This turned out to be a bunch of zombie lepers, back to avenge their deaths at the hands of the town's founders, who'd forced them to set sail and face certain death in a similarly thick fog. 'What you can't see won't hurt you...' proclaimed the trailer's husky voiceover, '...it will kill you.'

Just as the inhabitants of Antonio Bay could see the zombie-leper fog coming towards them, advection fog also forms as a result of moving air currents. It forms when low-level moist air drifts across a cooler surface. This usually happens at sea in spring and early summer, when air drifts from an area of warm sea surface to one where the surface is colder. Having picked up moisture from the warm surface, the air can form fog droplets as it cools. The association of advection fog with oceans is why it is often known as 'sea fog'.

Radiation fog, by contrast, never forms over water – only over land, usually on clear, calm nights. It is caused not by air drifting into a different environment, but by a stationary mass of air cooling as the ground below radiates away its heat. The air doesn't need to move much, but gentle motion can help the cooling process spread throughout the lowest layer, ensuring a good depth of fog.

The appearance of this form of terrestrial Stratus is closely linked to an absence of its airborne cloud cousins, making radiation fog somewhat the loner of the cloud world. This is because cloud cover overhead acts as a blanket, stopping ground temperatures dropping so low at night. The blanket re-radiates some of the Earth's warmth back down, reducing its nocturnal cooling. On clear nights, without it, the ground loses heat into space much more rapidly, creating ideal conditions for water vapour in the air at ground level to condense as fog, particularly during the autumn and winter, when the nights are long.

Radiation fog tends to clear when the wind picks up and disperses the layer by mixing it with drier air. But, on still mornings later in the year, it can lift as the Sun rises and warms the ground again. By picking up energy with the rest of the air, the water molecules begin deserting the fog's droplets faster than they join them. The droplets evaporate back into water vapour – water's invisible state, when it's flying around as individual molecules.

The lifting of a particularly deep layer of fog in this way can result in a low, but airborne, Stratus cloud being left above. When the lifting fog began as a shallow layer, it can remain for a time in a thin plane above the ground. I remember seeing a spectacular instance of this after a clear night in Australia. The radiation fog had developed in the early hours of the morning up to the height of my neck. As the ground warmed with the dawn, it cleared from the ground up to chest height. Walking through this suspended floor of cloud was weirdly disconcerting. I felt like a ghost pacing a house at the level of sunken corridors from centuries past. I was, of course, a rather solid phantom in a decidedly ethereal house.

Advection and radiation are the most common types of fog, but they are certainly not the only ones.

'Steam fog' appears when cold air flows over warm water (the opposite of advection fog) and the vapour evaporating off the water's surface instantly cools enough to form into droplets. The swirls of rising droplets are the evaporation process made visible, for water is constantly rising from the sea surface as vapour, but

RADIATION FOG

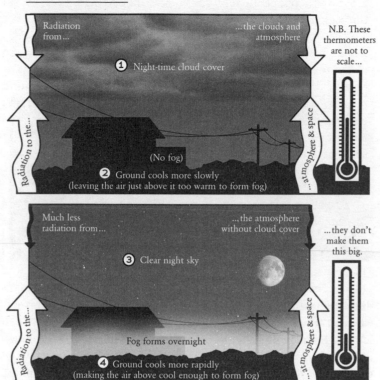

'Radiation fog' results from the air at ground level being cooled as the Earth loses heat rapidly on a clear night.

normally you can't see it. This type of fog is at its most dramatic in polar regions, where it is known as 'Arctic sea smoke'.

'Upslope fog' occurs when a gentle breeze blows moist air up the sides of a hill or mountain and the drop in pressure cools the air enough to form droplets. 'Valley fog' forms when air is cooled at night over higher ground and so becomes denser and sinks to lower ground. If the air cools enough then the fog can fill up the valleys, flattening the landscape with the glacier-like cloud.

'Freezing fog' is when temperatures are low enough for the water droplets to freeze into icy 'rime' on contact with solid objects. This is not to be confused with 'ice fog', where the fog's

Stephen Cook (member 132)

'Valley fog'. So named for obvious reasons.

droplets have frozen into crystals in the air. This only happens when air temperatures are extremely low – usually less than –30°C. Ice fog glitters beautifully in the sunlight. As with the larger, falling ice crystals of its close relative, 'diamond dust', it can interact with the sunlight to produce beautiful optical effects, which are similar to those found in the high ice clouds called Cirrus and Cirrostratus. Out of all the fogs, the rare, jewel-like sparkle of ice fog stands out as the most magical.

☁

I LOVE LOOKING at photographs of clouds almost as much as looking at the real things. They feel like the nearest that un-manipulated photography comes to abstract art – the nearest it gets to being a record of the world and an expression of a feeling at the same time.

The American photographer, Alfred Stieglitz, felt the same way. In 1922 he began taking an extended series of cloud photographs, which he later called 'Equivalents', and was the first photographer to undertake cloud photography solely for its artistic merits. The Equivalents were high-contrast black and white prints. For the first few years they included landscapes, but in 1925 he began photographing directly upwards so that the clouds completely filled the frame. He considered the shots as expressions of his state of mind. 'I have a vision of life,' he wrote in a letter to a friend, 'and I try to find equivalents for it.'[3]

Stieglitz was not just a photographer. He also ran 291, an art gallery on New York's Fifth Avenue, through which he became an influential champion of abstract art. He was instrumental in introducing America to the pre-war European *avant-garde* artists and, between the years 1908 and 1914, his 291 was the first US gallery to exhibit the emerging talents of Matisse, Rousseau, Cézanne and Picasso.

This passion for abstract art (something that Stieglitz described as 'a new medium of expression – the true medium'[4]) was only matched by his determination to see photography accepted as an art form in its own right. You might see the two as being in conflict: the art was *avant-garde* because it rejected realism, while photography, by its nature, could hardly have been more representational.

The Equivalents series of cloud photographs was how Stieglitz resolved this conflict. Clouds are Nature's abstract art – the moods of the sky – and they are the perfect subject-matter for expressing abstract emotion with photography. 'I know I have done something that has never been done. – May be an approach occasionally [found] in music,' Stieglitz wrote to a friend.[5] He sought 'through clouds to put down my philosophy of life – to show that my photographs were not due to subject-matter – not to special trees, or faces, or interiors, or special privileges, clouds were there for everyone – no tax as yet on them – free.'[6]

☁

IT'S THE END OF THE DAY in London and I'm on my way back home. The Stratus nebulosus opacus that had such a negative effect on my mood this morning has undergone a transformation.

David Fuller (member 62)

Stratus – at its best when viewed from the top of a mountain.

During the afternoon, following a fall of light drizzle, the formless grey layer developed a clumpy underside. Some parts of it brightened to white, whilst others darkened to a slate. The Stratus had changed into a clumpy layer, called Stratocumulus. It was a transitional cloud, forming as the layer began to clear.

I reach home with the setting Sun and gaps have appeared, through which I can see high delicate wisps of the ice-particle cloud, called Cirrus. Now the low cloud is in shadow, but the Cirrus above still catches the sunlight – its brilliant yellow streaks gleaming against the darkening winter sky.

I feel elated. With the clearing of the oppressive Stratus, I think of words by the American poet, James Russell Lowell:

> Who knows whither the clouds have fled?
> In the unscarred heaven they leave no wake,
> And the eyes forget the tears they have shed,
> The heart forgets its sorrow and ache...[7]

To a cloudspotter, the expanse of the heavens is all the more satisfying when it is revealed. How could I have bemoaned the only cloud that is prepared to visit us in the form of swirling mists and fogs? Would I have felt as euphoric had the sky looked this way all day long? No, *I wouldn't*. Stratus is like the magician's silk – whipped away just when we think all is lost, to reveal once more the spectacle of the sky.

FOUR

STRATOCUMULUS

The low, puffy layers

One moment clouds feel oppressive and smothering, and the next they are the very things that inspire us to dream. Who hasn't gazed up at castles in the sky and imagined a world away from the concerns of terra firma? As a Stratocumulus cloud develops out of a Stratus, glimpses of broken blue can begin to appear. Hours before, the Sun seemed smothered, but now the foggy layer has started to gather into snow-covered mountains and melt into winding rivers of blue. There is another world up there – a shifting terrain of glacial valleys and billowing peaks, a land of promise and escape – one with its own nebulous laws of geology.

The two main characters in Aristophanes' comedy, *Birds*, first performed in 414BC, were tired of their home city of Athens. Its tedious bureaucracy and incessant legal disputes were wearing the old men down and so, like many a modern-day city dweller, they decided to move out of town for a more peaceful life. Leaving the city and their debts behind, they went in search of Tereus, a character from Greek myth, whom the gods had turned into a hoopoe bird. Their reasoning was that Tereus, having once been a human and now a bird, might have come across 'a nice cushy city, soft as a woollen blanket, where we could curl up'.[1]

When they found him, however, none of the bird's suggestions were quite the ticket. Cities just aren't carefree places, they realised, and whilst the life of birds is carefree, they don't live in cities. Then

STRATOCUMULUS CLOUDS

Stratocumulus are low layers or patches of cloud, with well-defined bases. They are usually composed of clumps or rolls, and often show strong variations in tone – from bright white to dark grey. Their cloud elements may be joined into continuous, unbroken layers or have gaps between them.

TYPICAL ALTITUDES*:
2,000–6,500ft
WHERE THEY FORM:
Worldwide – it's a very common cloud.
PRECIPITATION (REACHING GROUND):
Occasionally light rain, snow or snow pellets.

Blue sky showing through

Stratocumulus stratiformis opacus...   ...and perlucidus

STRATOCUMULUS SPECIES:

STRATIFORMIS: The most common, when the clumps or rolls extend over a large area. A 'roll cloud' is a particular formation, in the shape of a large, individual tube of cloud.

LENTICULARIS: When one or more mass of cloud is in a smooth, solid-looking almond or lens shape.

CASTELLANUS: When the elements have crenellated tops.

STRATOCUMULUS VARIETIES:

OPACUS: When the layer is thick enough to completely mask the Sun or Moon.

TRANSLUCIDUS: When it is thin enough to show the outline of the Sun or Moon.

PERLUCIDUS: When there are gaps between the cloud elements.

DUPLICATUS: When there are layers at different altitudes, sometimes partly merged.

UNDULATUS: When the elements are arranged in nearly parallel lines.

RADIATUS: When lines of closely bunched elements appear to converge towards the horizon.

LACUNOSUS: When the layer shows large net-like holes fringed with cloud.

NOT TO BE CONFUSED WITH...

CUMULUS: which is also clumpy, well defined, and forms at similar altitudes. The elements of Stratocumulus tend to be closer together and to have flatter tops.

ALTOCUMULUS: which is a mid-level layer of cloudlets. These appear smaller than the Stratocumulus elements, which – looking above 30° from the horizon – appear larger than the width of three fingers, held at arm's length.

STRATUS: which is a low, indistinct layer, with much less variation in tone and less definition than Stratocumulus.

* These approximate altitudes (above the surface) are for mid-latitude regions.

one of the old fellows proposed an idea to the birds. What if together they founded a city in the sky – one up in the clouds, away from earthly concerns? This new city would become all-powerful, he explained, for the birds could hold the gods to ransom: it would be up to them whether or not to let the *The ultimate* fumes from sacrifices below pass up to Zeus and the others *escape* above. It would also be the perfect place for the two Athenians to escape to. So delighted were the birds with the old man's plan that they immediately agreed and appointed him as their leader.

Eating a magical root, he was able to grow wings, which was clearly going to be a help in navigating the new city. Nevertheless, escaping to a Utopia in the clouds was not without its difficulties. Not only did the new leader of the birds and his friend have to fend off a whole host of profiteers and tricksters, who wanted to come and live in their cloud city, they also had to contend with some rather angry Olympian gods.

It all turned out well in the end, though. With a little persuasion, they managed to convince the gods to hand power to the birds, and pretty soon the old fellows ruled the roost in their dream city in the sky. Nevertheless, you might think that anyone who dreamt of doing the same was living in Cloudcuckooland. You wouldn't be far wrong, for that term is actually a translation of the Greek word 'Nephelokokkygia' – the very name that the two Athenians chose for their fantastical Utopian city.

What is a cloudspotter to do while waiting for a Stratus to change into a Stratocumulus and break to reveal the heavens? Look up, of course, and dream of escape to their own Cloudcuckooland.

☁

STRATOCUMULUS IS A LOW layer of cloud, usually forming between 2,000 and 6,500ft in temperate regions, which consists of clumps and mounds. These are often Cumulus-like in appearance and can be joined together into a continuous layer or have some gaps between. Either way, the layer shows much more variation in its tones than the Stratus and tends to have a clearly defined texture to its base. Its shades can vary from bright white to a dark, bluish grey. Though not normally associated with much precipitation, the Stratocumulus can, when its mounds grow tall enough, produce light rain or snow. If heavy showers appear to fall from one, this is usually because a Cumulus congestus or even a Cumulonimbus is 'embedded' in the layer – its tower rising up above it, hidden from view below.

The Stratocumulus can be considered half-way between the individual, free-floating Cumulus and the formless layer of the Stratus. And of all the low clouds, it stands out as exhibiting the greatest variation in its appearance.

The different species and varieties of this genus depend mostly on the shape and arrangement of its elements. There are three recognised species: stratiformis, the most common, where the clumpy layer extends over much of the sky, rather than being just in one or more isolated patches; castellanus, where the separate elements of the cloud layer have turret-like crenellations above their smoother bases; and lenticularis, where they have smooth lens or almond shapes (sometimes this species appears less like a layer of clumps, and more like one individual almond-shaped cloud). As with any of the cloud genera, a Stratocumulus does not *have* to be one of these recognised species: if it doesn't fit one of the above descriptions, a patch of low clumpy clouds is just called Stratocumulus.

Cloudspotters who want to get the naming right should remember that examples of any genus of cloud can only be considered to be of one species at any time. They can, however, exhibit any combination of varieties, which are characteristics of appearance common to that cloud type. Being a cloud of such variation, the Stratocumulus is the proud possessor of seven recognised variations:

1) Duplicatus: when it is in more than one distinct layer, these being at different altitudes.

2) Perlucidus: where the cloud clumps have gaps between them, through which sky or higher cloud is visible.

3) Lacunosus: a rare variety, rather like the inverse of perlucidus, with larger holes separated by loose, honeycomb-like rivers of cloud between.

Stratocumulus – looks like someone couldn't find the 'off' switch
on the candyfloss machine.

4) Radiatus: a variety shared with Cumulus clouds, when the layer's clumps are in more or less parallel lines that extend far enough into the distance for them to appear to converge towards the horizon.

5) Opacus: which, like the following two varieties, is also used to describe Stratus. It is when the layer is thick enough to obscure the position of the Sun or Moon.

6) Translucidus: when, by contrast, the layer is thin enough for you to be able to discern their positions (translucidus and opacus are the only variations that are mutually exclusive).

7) Undulatus: when the layer is in the form of parallel lines of cloud – either distinct rolls, with gaps between, or ones that merge together, so that the layer's base has parallel, wave-like undulations.

With so many different guises, the Stratocumulus is always in transition. It is not unlike the pop singer, Cher, at the height of her costume-changing stage routines – always nipping off stage to reappear in a more fantastical outfit. And how varied the Stratocumulus's garments can be! One of the more dramatic is known as a 'roll cloud'. This is when the Stratocumulus is in the shape of a long tube – sometimes with a smooth, glacial surface, other times with a puffy one, like a very elongated Cumulus.

The Cher of the cloud family

Don't laugh, but I once flew all the way across the world just to see a Stratocumulus roll cloud. Called the Morning Glory, it forms during the spring months of September and October in northern Queensland, Australia.

Often longer than Britain itself, this enormous cloud passes over the coast of the Gulf of Carpentaria. It forms in the middle of an enormous wave of air, travelling at speeds of up to 40mph, and it has made the tiny outback settlement where it forms a Mecca for glider pilots, who surf the cloud like a wave. So unique and dramatic is this particular Stratocumulus formation that I've given it a chapter of its own (see page 283). No doubt, the Morning Glory would be Cher in the brass armour bikini and gold Viking helmet outfit she wore on the sleeve of her 1979 album, *Take Me Home*.

It may seem ridiculous that the same Stratocumulus classification should apply to clouds as different in appearance as the long, smooth tube of the Morning Glory roll cloud and the clumpy, extended layer of individual cloud heaps called Strato-cumulus stratiformis perlucidus. The two seem to be worlds apart.

But cloudspotters should remember that the Stratocumulus genus is used to refer to any low clouds (below 6,500ft) that are neither individual convection clouds, such as the various species of Cumulus, nor misty, featureless layers like the Stratus. It is a catch-all category for low clouds that refuse to toe the party line – all those more free-thinking ones that flout the customs we use to categorise their low-level cousins.

Of course, clouds pay little attention to the rules of behaviour we presumptuously ascribe to them. Chaotic to their misty core, they do their best to confound our attempts at classification.

How can a body so nebulous, ephemeral and mutable ever be pigeon-holed? Cloudspotters will come to love the cloud's rebelliousness – just when they think they have identified a formation, it will change and mock their attempt to pin it down.

When Luke Howard proposed his cloud classification system during his seminal 1802 lecture, *On the Modifications of Clouds*, part of his genius was to appreciate the futility of considering clouds as fixed forms. They are in a constant state of transformation – something that photographs and paintings can never convey. By introducing the terms Cumulus, Stratus, Cirrus and Nimbus, Howard was referring to passing moments in the constant transformation of cloud shapes.

As water in transition, clouds can be momentary stages in the incessant cycle of rising and falling water – like a ball hanging gracefully in the air at the apex of a tennis player's lob.

☁

ONE WAY THAT Stratocumulus form is by the spreading and joining of Cumulus. Cloudspotters might be curious to know what tectonic shifts of the atmosphere cause free-floating islands of Cumulus to gather into the hilly aerial terrain of the Stratocumulus. Well, it tends to happen when the Cumulus have formed under what meteorologists call an 'inversion'.

You can't see an inversion, because it is related to the air temperature, and consequently it is not the easiest thing to get your head around. But any cloudspotter would be wise to spend a moment or two understanding it, for it plays an important role in the way many clouds spread out, not just the Stratocumulus.

The troposphere is characterised by the air generally becoming colder with altitude. This is the case on average, but it is not always so. The movement of warm and cold air around the globe and the way heat is transferred to and from parts of the atmosphere during the day and night sometimes causes a mass of warmer air to find itself above a cooler one. In such a region, the air becomes briefly warmer again with height. For a while there is an inversion of the normal temperature pattern. This can happen at any altitude in the troposphere, either over localised regions or across thousands of square miles.

The reason an inversion is relevant to cloud formation is that it can act as an invisible ceiling to their vertical growth. On encountering an inversion, the rising thermals forming Cumulus clouds can all of a sudden find themselves no longer warmer and lighter than the air around. All they can do is spread out sideways. Below the inversion, the clouds join, like plumes of smoke gathering at the top of a greenhouse while the pipe-puffing gardener tends his tomatoes below.

The same principle explains the spreading anvil of the Cumulonimbus storm cloud. In the case of that mountain of a cloud, the invisible ceiling of temperature inversion is usually the 'tropopause', which is the top of the troposphere, where temperatures begin to stop decreasing with altitude (a result of the way the gases, such as ozone, in the lower stratosphere absorb the Sun's ultraviolet rays).

In the case of the lower, fair-weather Cumulus clouds, if their rising thermals are not energetic enough to burst through a localised temperature inversion, they can expand laterally, merging into a layer of joined-up clumps – a layer of Stratocumulus.

☁

WE ALL NEED TO ESCAPE every once in a while. I'm not talking about queuing up at airports and crowding on to planes to burn on the beach with the other package-holiday tourists. For a cloudspotter, there is a form of escape that is much closer to home – one that costs nothing and is guaranteed to benefit the soul.

I call it 'contemplating the heavens below' and its effectiveness depends greatly on adopting the right frame of mind and assuming the appropriate posture. Cloudspotters should find a place with some elevation – a hill, perhaps, or in front of an upstairs window – and lie on their back so that they can look upwards and behind them to the clouds. A sky with a dramatic cloudscape of Stratocumulus is a good one with which to try this.

Looking at the clouds, they should allow a shift in perspective

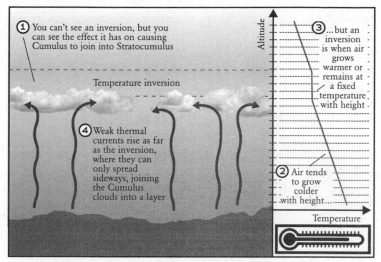

① You can't see an inversion, but you can see the effect it has on causing Cumulus to join into Stratocumulus

Temperature inversion

④ Weak thermal currents rise as far as the inversion, where they can only spread sideways, joining the Cumulus clouds into a layer

Altitude

③ ...but an inversion is when air grows warmer or remains at a fixed temperature with height

② Air tends to grow colder with height...

Temperature

A 'temperature inversion' is one of the reasons why Cumulus spread and join to form a puffy layer of Stratocumulus.

to take place. From this vantage-point, it is as if they are no longer looking up at the clouds, but gazing down on them from on high. They are suspended above a fantastical cloudy land, stretching into the distance.

Cloudspotters should take the trouble to map this strange land, for it will never be seen again. They should survey the contours of its terrain – chart its gentle undulations, trace its winding valleys, pause on its dark summits. It is a land where the light behaves differently from normal – it shines from the valleys, casts shadow on the peaks. In fact, the terrain glows from within.

Cloud meditation

The problem of what music to listen to while contemplating the clouds has finally been solved. Nicolas Reeves, a professor at the University of Quebec in Montreal, Canada, has invented the Cloud Harp – an instrument that creates music from the shape of the clouds above it. So far, it has played in six cities around the world – Amos and Montreal (Canada), Lyon (France), Hamburg (Germany), Gizycko (Poland) and Pittsburgh (USA).

When the sky is blue, the harp is silent, but with the first appearance of clouds above it, the music begins. 'It uses a lidar,' explains Reeves, 'which is a laser beam directed at the clouds.

A large 'Cloud Harp', installed at the Society for Arts and Technology, Montreal, in 2004. The harp's music is triggered by a 'lidar', which measures changes in the cloud cover above.

Whatever bounces back to the instrument is measured and gives us an idea of the brightness of the cloud, as well as its height.' A musician, known as a 'cloudist', configures the instrument so that this information triggers and controls particular musical sounds. He then leaves it to play the music of the clouds to passers-by.

Sometimes musicians are brought in to provide orchestrations by recording samples for the harp to play. 'It means that we can play the clouds of St Louis, Missouri, through an arrangement by Trillian Bartel from Hamburg,' explains Reeves.

In Amos, Northern Quebec, the Cloud Harp was installed in a park clearing, surrounded by trees. 'When there was a full Moon,' he remembers, 'people brought their sleeping bags and stayed the night next to the harp. They just lay there listening to the clouds – it was fantastic.'

Contemplating the heavens below, a cloudspotter can always escape – even if for just a few minutes – from the trials and pressures of life. Let others dream of escaping to a place in the Sun. Cloudspotters know better. They can visit the world that the American author and naturalist, Henry David Thoreau, observed in the light of the dying day:

> *Between two stupendous mountains of the low stratum under the evening red, clothed in slightly roseaceous amber light, through a magnificent gorge, far, far away, as perchance may occur in pictures of the Spanish coast viewed from the Mediterranean, I see a city, the eternal city of the west, the phantom city, in whose streets no traveller has trod, over whose pavements the horses of the sun have already hurried, some Salamanca of the imagination.*[2]

◠

IN THE CLOSING SEQUENCE of Steven Spielberg's *Close Encounters of the Third Kind*, an enormous UFO descends on to Devil's Tower in Wyoming, and Roy Neary, the character played by Richard Dreyfuss, climbs on board with an assortment of American scientists to be taken off to who knows where.

I might have just ruined the film for anyone who still hasn't seen it, but this is in a worthy cause, for the special effects that appear moments before the mothership's arrival help to explain how a temperature inversion can lead to Stratocumulus clouds.

Moments before it appears, a host of smaller craft descend in an advance party. They emerge from a layer of thick, billowing clouds that spread out in time-lapse fashion across the desert sky. These were the first convincing special-effect clouds in the history of cinema, and they were created by the visual effects pioneer Douglas Trumbull. Though they swelled down behind the *Close Encounters of the* craft in a somewhat un-meteorological fashion, they *of the* spread out in much the same way as real Stratocumulus *Stratocumulus Kind* do. In order to create them, Trumbull had to invent and build a piece of equipment specifically for the task. The device became known as a 'cloud tank'. It revolutionised cloud special effects and it worked on the principle of a temperature inversion.

Trumbull's clouds didn't consist of water droplets suspended in the air, like the real thing, but tiny globules of paint in a tank of water. To have the necessary control over their behaviour and to be able to light them effectively, he knew that his clouds would need to be miniature ones. 'I had an idea that realistic miniature clouds could be created in a liquid environment – into which would be injected some other milky white liquid,' he said in a 1977 article in *American Cinematographer* magazine. So he built a seven-foot-square glass tank at his special-effects studio, into which a remote-controlled arm could be lowered to inject a special mixture of white poster paints.

The tank was like a large aquarium – only one for someone who couldn't decide if they wanted fish of the cold-water or tropical variety. The bottom half of the tank was filled with cool water and

Ian Watterson (member 1524)

To be a Stratocumulus, a cloud layer has to be below 6,500ft. It is hard to judge the altitude in this photograph, but the undulatus billows are so pleasing that it is going here anyway.

the top half with warm. Of course, the water temperature would always have a tendency to even itself out. But a complicated system of plumbing, heating and filtration allowed the effects co-ordinators to sustain a temperature inversion in the water, in which a layer of dense cold water below was covered by one of less dense warm water above. In the boundary region between the two layers, the remote-controlled arm was carefully manipulated to squirt its clouds of white paint. These were lit from above to simulate moonlight, and had fibre-optic probes lowered into them to create lightning effects. Spielberg could then point the camera up through the water and film the clouds from below.

The temperature of the paint solution was half-way between that of the warm and cool water, which meant that it was also half-way between the two in density. When injected into the boundary layer, the paint swelled upwards only as far as the ceiling of warm water above and sank down only as far as the floor of cold water below. And just as we can't see a temperature inversion in the air, the inversion in the water was not visible to the camera. Trumbull's clouds gathered and spread in a puffy layer in the same way as Stratocumulus do below a temperature inversion.

'Since each "take" required a totally fresh and clean tank of specially heated (or cooled) and filtered water,' explained Trumbull, 'the shooting was slow and difficult and occurred on and off for over a year to achieve the results we wanted.'

It may have been a painstaking process to make puffs of white paint in a schizophrenic fish tank look like clouds across the Wyoming sky, but Trumbull's work on *Close Encounters* earned him a nomination for Best Visual Effects in the 1978 Oscars. I might be biased in saying it was a scandal that he didn't win, but surely his fine work in creating credible clouds should at least have won him 'Best Use of a Temperature Inversion in a Supporting Role'?

☁

A FEW YEARS BACK, I went to see an art exhibition at London's Tate Britain called *American Sublime: Landscape Painting in the United States 1820–1880*. Room after room showed huge canvases by the great painters of nineteenth-century America. They were dramatic scenes of idealised landscapes – wild, untamed prairies, lakes and mountains stretching as far as the eye could see, which the catalogue explained reflected the expansionist, frontier spirit of the young continent's European settlers. My interest, of course, centred on what was happening in the skies.

As I looked at paintings such as *Twilight in the Wilderness* by Frederic Edwin Church and *Storm in the Rocky Mountains – Mt Rosalie* by Albert Bierstadt, I was struck by the way their fantastical cloudscapes mirrored the drama below. At times, the spectacular skies seemed even more of a celebration of sublime wilderness than the landscapes themselves. The parts of the paintings above the horizon expressed the frontier spirit of the age even more than those below it.

Feeling the urge to perform the gallery-goer's version of 'contemplating the heavens below', I bought the catalogue to see what the paintings would look like upside down. Of course, the

Churches and Bierstadts would have been turning in their graves, but it was an enlightening experiment.

Looking at these sublime scenes with the clouds serving as the landscapes, the effect was really not so different from when they were the right way up. I couldn't help wondering how long it *Upside-down* would have taken the public to notice if one of the *paintings* dramatic canvases had accidentally been hung upside down. What if the hyper-real vermilion clouds of Frederic Edwin Church's *Sunset, Bar Harbor* appeared as the landscape and its dark, hilly silhouette of land as the brooding clouds above?

The inverted painting seemed to me to embody just as effectively the new republic's ambitions for the further reaches of its territories. Being clouds that generally form several miles above the Stratocumulus, the inverted sunset-tinged Altostratus and Cirrus gave the impression I was looking at their grand vista from way above.

Lost in these terrains of the imagination – the very same that a cloudspotter explores lying on the ground – I was embarrassed to realise people were staring at me. What a philistine I must have looked, standing in the august galleries of Tate Britain leafing through the catalogue upside down.

☁

CUMULUS SPREADING below an inversion is not the only way that Stratocumulus clouds form. Another is when they develop from a stable, flat Stratus – when that calm, foggy layer of cloud gathers into clumps.

What causes a low-hanging blanket of Stratus to get roughed up like this? Winds picking up and creating turbulence at the cloud level is one cause. Another is when the Stratus layer is thin enough to let enough sunlight through for gentle thermals to rise from the ground and stir it up. But Stratus can also bunch up and gather into Stratocumulus when there is little wind and the cloud is too thick to allow thermals to develop.

In this case, the transformation results from the way clouds absorb and shed heat. Like that business with temperature

Frederic Edwin Church's *Sunset, Bar Harbor* (1854). Would anyone have noticed
if it had been hung upside down?

inversions, it is something any cloudspotter would be wise to get
the hang of, for it is an important factor in the formation of clouds
in general, not just the Stratocumulus.

There are four ways in which heat travels and all of them
influence the formation of clouds. 'Convection' is the way warm
air carries heat as it floats upwards in a thermal. The lava lamp
example showed how heat can travel around like this by the
movement of liquids as well as gases. 'Conduction' is how heat
travels between things that are touching, or along the length of
something, as a result of the hyperactive molecules of the warmer
part agitating their calmer neighbours until they are all jiggling at
similar speeds. You see it when a snowball melts in your hand. It is
also how, after a clear night, air touching the cold ground loses heat
into the earth and cools enough to form fog. 'Vaporisation' is how
sweat evaporating off your skin cools you down – the water takes
heat from your skin as it changes into a gas. It is also the way that
the air warms slightly when the droplets of a cloud first appear –
something that causes the air to rise in the cauliflower mounds of
a Cumulus as it expands and floats due to the heat released when
the droplets form. And then there is 'radiation'.

This is beginning to sound like a physics lesson, but it would be
a mistake to play truant. Cloudspotters should come back from

Nick Lightbody (member 95)

The ragged terrain of Stratocumulus obeys its own laws of aerial geology.

behind the bike sheds and spare a few moments for radiation. It is the most important of the four, you see. Not because of its role in encouraging a Stratus to change into a Stratocumulus, of course, but because without it the Earth would be too cold to support life. Radiation is how heat reaches us from the Sun through the vacuum of space. It is rather different from the other three, since it takes the form of 'electromagnetic waves'.

Of all the electromagnetic energy radiated from the Sun, the portion we can see as visible light is only a very narrow band of wavelengths. It nevertheless accounts for around 45 per cent of all the energy the Sun emits. Nine per cent falls in the range of shorter wavelengths – such as 'ultraviolet' radiation, which we cannot see but causes sunburn – and the remaining 46 per cent is spread across a broad range of longer wavelengths called 'infrared', which we also cannot see but can feel as warmth. All objects emit radiation, and the hotter they are, the shorter the wavelength of electromagnetic waves they radiate most intensely. (This is why as objects become hotter they first glow red, then yellow, and eventually blue.) The Earth, being much colder than the Sun, emits radiation most intensely in the long, invisible, infrared wavelengths.

Whilst all objects emit electromagnetic radiation, they can be quite specific about what wavelengths of radiation they absorb. This depends on the particular types of atoms or molecules that they consist of. Cloud water droplets tend to absorb more of the longer, infrared wavelengths of radiation, merely reflecting *Wake up at* away much of the shorter visible and ultraviolet light. This *the back!* is why cloud cover leads to cooler days (by reflecting much of the sunlight's predominantly shorter wavelength radiation back up, without being warmed in the process) and leads to warmer nights (by absorbing much of the Earth's longer, infrared, wavelength radiation and emitting some of it back down again).

It may have taken a while, but the time has finally come to explain how a flat Stratus cloud can transform into a clumpy Stratocumulus without the influence of external winds or thermals.

The Stratus cools at its top and warms at its base. The top of the cloud layer absorbs little of the Sun's short-wave radiation from above – merely reflecting it away – but the bottom absorbs a good deal of the Earth's long-wave radiation from below. Being warm below and cold above is an unstable state of affairs for any cloud to be in, since the lower warm air expands and starts to float *This week's* upwards through the colder, denser air above. Once a calm, *homework:* flat, stable layer of Stratus, the cloud begins to develop *Look Up* eddies of convection as some parts of the warmer air float up. These churn the layer, causing some regions to build and others to thin out. Enter the Stratocumulus.

That wasn't so painful, was it?

☁

OF ALL THE WEIRD and fanciful lands visited by Lemuel Gulliver in *Gulliver's Travels*, the eighteenth-century satirical novel by Jonathan Swift, Lilliput is the one that is best remembered on account of its minuscule inhabitants with grandiose ideas. But the people on one of the other islands encountered by Swift's intrepid traveller are of particular interest to a cloudspotter.

When Gulliver chanced upon the island of Laputa, he was surprised to see that it was floating above the clouds. It was held aloft by the action of an enormous magnet – a 'Load-stone' –

Azhy Chato Hasan (member 1687)

What is wrong with having your head in the clouds?

contained within its base. The Laputans were able to make their floating island move over their king's domain by changing the stone's orientation. They were, it turned out, a rather weird bunch:

> *Their heads were all reclined, either to the right, or the left; one of their eyes turned inward, and the other directly up to the zenith.*

Their outward garments were adorned with the figures of suns,
moons, and stars.[3]

Living up in the clouds, the Laputans were also a somewhat
distracted race – their minds being always occupied with obscure
mathematics and music. One of the first things that Gulliver
noticed about the Laputans was their attendant 'flappers'. These
servants carried a bladder containing pebbles, fastened to the end
of a short stick, with which they thumped their master's ears when
someone was addressing them, and their mouths when they were
expected to reply:

> *It seems the minds of these people are so taken up with intense*
> *speculations, that they neither can speak, nor attend to the discourses*
> *of others, without being roused by some external taction upon the*
> *organs of speech and hearing.*[4]

Gulliver didn't exactly take to these queer folk. Though he
respected their mathematical and musical talents, they were so lost
in their own worlds that he could hardly relate to them at all.

When cloudspotters let their minds wander up to the shifting
terrains of the heavens, they will often be told that they, like the
Laputans, are living with their head in the clouds.

They should pay no attention – of course they are. And what's
so wrong about that?

The Middle Clouds

FIVE

ALTOCUMULUS

The layers of bread rolls in the sky

Altocumulus is a 'mid-level' cloud, usually comprising a patch or layer of more or less regularly spaced cloud clumps, which are often given the agreeable name 'cloudlets'. The cloud forms around half-way between the ground and the top of the troposphere.

It may seem somewhat confusing that mid-level clouds should have the prefix 'Alto-', meaning high in Latin. Wouldn't 'Medio-' have made more sense? Perhaps, but it was nevertheless proposed by Émilien Renou, the director of the French observatories at Parc Saint-Maur and Montsouris, in 1855 to signify clouds at this level and was subsequently accepted by the meteorological community in the 1870s.

The altitude of the middle region of the troposphere is trickier to define than you might think, for the height of the troposphere depends greatly on where you are. The atmosphere as a whole reaches higher in the tropics than it does at the poles, largely because it expands more over the warmer surfaces around the equator than elsewhere. Consequently, the troposphere often extends to around 60,000ft (over 11 miles) in the tropics, but only around 25,000ft above the poles.

The variation in troposphere height with latitude means that there is no fixed range of altitudes in which the mid-level clouds form. To keep things simple, I'll stick to the temperate, middle

HOW TO SPOT
ALTOCUMULUS CLOUDS

Altocumulus are mid-level layers or patches of cloudlets, in the shape of rounded clumps, rolls or almonds/lenses. These are white or grey, and the sides away from the Sun are shaded. Altocumulus are usually composed of droplets, but may also contain ice crystals.

TYPICAL ALTITUDES*:
6,500–18,000ft
WHERE THEY FORM:
Worldwide.
**PRECIPITATION
(REACHING GROUND):**
Very occasionally causes light rain.

ALTOCUMULUS SPECIES:
STRATIFORMIS: Most common, when the cloudlets extend over a large area.
LENTICULARIS: When it is in the form of one or more individual almond- or lens-shaped masses that appear dense, with pronounced shading.
CASTELLANUS: When the cloudlets have crenellated tops.
FLOCCUS: When the cloudlets are Cumulus-like tufts, with ragged bases, often with fibrous trails (virga) of ice crystals falling below.

Altocumulus stratiformis undulatus

Altocumulus lenticularis

Altocumulus floccus

ALTOCUMULUS VARIETIES:
OPACUS: When the layer is thick enough to completely mask the Sun or Moon.
TRANSLUCIDUS: When it is thin enough to show the outline of the Sun or Moon.
PERLUCIDUS: When there are gaps between the cloudlets.
DUPLICATUS: When there are layers at different altitudes, sometimes partly merged.
UNDULATUS: When the cloudlets are arranged in nearly parallel lines.
RADIATUS: When long lines of them appear to converge towards the horizon.
LACUNOSUS: When the layer shows net-like holes fringed with cloud.

NOT TO BE CONFUSED WITH...
CIRROCUMULUS: which is a higher layer of cloudlets, that appear like little grains of salt. Looking above 30° from the horizon, the larger Altocumulus cloudlets generally appear the width of between one and three fingers, held at arm's length. Also, these exhibit shading, which those of Cirrocumulus don't.
CIRRUS: which is a high cloud, whose streaks of falling ice crystals can resemble Altocumulus cloudlets showing virga, but do not have their dense-looking heads.

* These approximate altitudes (above the surface) are for mid-latitude regions.

Top: Stephen Cook (member 132); Below left: Mike Cook (member 1690)
Below right: Terry Falco (member 1592)

latitudes where the troposphere reaches to around 45,000ft. Here, we can say that Altocumulus and the other mid-level clouds form between 6,500 and 23,000ft up.

This is generally above the influence of thermals – the localised currents of air rising off the Sun-warmed ground. Thermals therefore do not have the central role in the formation of Altocumulus that they do for Cumulus. Cloudspotters should remember that the names of the genera, species and varieties of clouds depend more on the way they look and their typical altitudes than on how they form. Don't be confused by the formality of the Latin names. 'Altocumulus' just means that they are mid-level clouds that happen to be in a clumpy shape. It doesn't mean that they form in the same way as Cumulus, which is just as well, since they don't.

☁

ON 27 JULY 1907, in the little Norwegian town of Drøbak, twenty miles south of Oslo, someone took a photograph of the pleasant view across the Oslo fjord towards the town of Holmsbu on the opposite bank. The grainy black and white picture shows a couple of jetties in the foreground and some clipper ships anchored in deeper water. It also shows a dark disc hovering in the sky above them. Sixty years after the photograph was taken, it was published in the Italian Sunday supplement, *La Domenica del Corriere*, where it was held up as one of the earliest UFO photographs. 'Even today,' wrote the caption, 'the phenomenon is a mystery.'

Whilst it may have been an unidentified flying object, this was no flying saucer. It was in fact a particular species of Altocumulus cloud that is known as lenticularis. Although you can't see much from the photo – just a disc-shaped shadow – the clue to it being a cloud comes from the hill behind which it is hovering.

Is this the oldest UFO photo, or is it the work of a cloudspotter?

I said that Altocumulus is usually a layer or patch of fairly regularly spaced cloudlets, so it might seem odd to identify such an individual cloud as being part of this genus. It doesn't look very layer-like. The lenticularis does indeed appear quite different from the typical formations of Altocumulus, and has more in common with the lenticularis species of the lower Stratocumulus cloud.

In both cases, lenticularis are 'orographic clouds', which means they form when air is forced upwards as it passes over an obstacle such as a hill or mountain. They are fairly common in mountainous regions. Nevertheless, it is such a dramatic cloud that it feels like a special event when you come across one. Who knows if the anonymous turn-of-the-century photographer was a cloudspotter or if he just captured the Altocumulus lenticularis by accident? I like to think that he excitedly set up his tripod and bellows, and rushed to adjust the framing simply to record the cloud. Altocumulus lenticularis is, after all, one of the most dramatic and beautiful species there is.

If it looks more solid than many, that's because, like a young puffy Cumulus cloud, it is composed of a large number of very small droplets. The smaller and more plentiful a cloud's droplets, the more opaque it appears. But the 'lennie', as glider pilots tend to call it, has a much smoother, silkier surface than the crisp mounds of the convection clouds.

Lenticularis means lens-shaped. The cloud can look like a very elongated lozenge or sometimes like a stack of pancakes, but the classic shape is of a flying saucer. Any cloudspotter lucky enough *Aliens* to catch sight of one when snowboarding in the Alps might *stopping for* wonder if aliens have parked their spaceship in the lee of *Glühwein* the Matterhorn for a mug of Glühwein before the long ride home through the Milky Way. Of course they haven't. They've just come to remind us that the clouds are Nature's poetry, spoken in a whisper in the rarefied air between crest and crag.

☁

I WAS OVERJOYED to see the same unidentified flying lennies one summer while on holiday in Arezzo, Tuscany. They weren't

John Lamb (member 1478)

ABOVE: Altocumulus lenticularis in the lee of Mount Cook, New Zealand.
RIGHT: The same clouds in the skies of Piero della Francesca's fresco on the walls of the basilica in Arezzo, Italy.

in the sky above me, but frescoed to the walls of the Basilica di San Francesco – hovering in the skies of Piero della Francesca's fifteenth-century masterpiece, *The Legend of the True Cross.*

The sequence of frescoes tells the story of the wood used to fashion Jesus's crucifix. Most people rate them as one of the best depictions of the legend – how the wood came from a tree that grew from Adam's grave and how, after being hidden by King Solomon who prophesied its gruesome destiny, it was found, used for the crucifixion and subsequently fought over by emperors and kings. I, however, like the frescoes for their clouds.

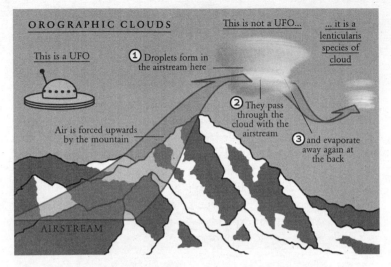

'Orographic clouds' form when air cools as it is forced to rise in order
to pass over an obstacle such as a mountain.

For some reason, della Francesca chose to populate his azure skies with Altocumulus lenticularis. While other fresco painters of the time were quite happy to use common old Cumulus, these clearly weren't good enough for Piero – he wanted fancy clouds. But why did he choose lennies – the same type that would later fool *La Domenica del Corriere* into publishing the 'first UFO photograph'? Perhaps his growing up in Borgo San Sepolcro (now called Sansepolcro) had something to do with it.

Situated close to the region of Umbria, at the foot of the Apennine Mountains, della Francesca's home-town would have been as good a place as any to see orographic clouds like the Altocumulus lenticularis. Did the artist look up one day as a young boy to see a squadron of beautiful lennies hovering in the lee of the mountains before him? Could della Francesca have had a moment of cloudspotting epiphany – one which would dictate his choice of clouds as an artist in later life?

Yes, of course he could.

I DON'T WANT TO devote too much attention to the lenticularis species, as there are many other wonderful types of Altocumulus to consider, but I do want to spend a little longer with it. As an orographic cloud, it does demonstrate one of the main ways that clouds form.

The convection clouds, like Cumulus, form when air rises in thermals from the Sun-warmed ground. Layer clouds, like Stratus, often form due to the gentle large-scale ascent of moist air when a region of warmer air is lifted as it comes into contact with a cooler one. But orographic clouds, such as lennies, form when winds encounter an obstacle, like a hill or mountain, and are forced upwards to pass over. Each type of cloud formation involves air rising. When air rises it expands, causing it to cool. By cooling, the air's molecules slow down and some of the water ones – the water vapour – end up joining together into droplets or even, if it's cold enough, ice crystals.

Suppose some cloudspotters are driving up a mountain for the view. They might feel their ears pop with the drop in pressure. An air stream rising up the side of the mountain will also move into an environment of continually lowering pressure.

When they stop the car to lower the tyre pressures, so that the wheels grip better on the snowy road, the cloudspotters might feel the nozzle of the tyres become colder as the air expands and rushes out. The stream of air flowing up the mountain and dropping in pressure also cools as it expands.

As they stand triumphantly on the mountaintop, the cloudspotters might notice puffs of mist forming in their breath as it cools by mixing with the air around. If the air stream that is riding up the mountain contains enough water vapour and cools enough by its gain in altitude, some of the vapour can also condense into cloud droplets, forming into orographic clouds that the spotter stands admiring.

Whilst this is the general principle of orographic cloud formation, the particular shape of the lenticularis species results when the air stream takes on a wave-like motion in the lee of the summit. It is much the same as the standing wave that can be seen when the current in a fast-moving stream flows over a large rock.

The surface of the water can also show a stationary wave shape downstream from the obstacle. Even though the water is rushing through, these crests of wave are stationary.

Exactly the same thing can happen in the airflow behind mountains and hills, and the stationary crests of the air current can be much higher than the mountain itself. Under the right conditions, lens-shaped clouds can form at each of the crests. A keen cloudspotter will see that, unlike most clumps of cloud, the Altocumulus lenticularis remains remarkably stationary even though there is often considerable wind.

The air is actually blowing through the cloud, forming droplets of water at the front of the crest that pass through with the air stream and then evaporate again as the air comes down the back of the crest. Though the droplets are speeding through the cloud, while the air flows at a constant speed, the point at which they form and they evaporate is fixed. Thus the shape of the cloud as a whole doesn't move.

Despite the rush of air passing through them, lennies find their parking place in the lee of mountains and hang there – worried, no doubt, that if they move they'll never find such a good parking space again.

☁

THOUGH THEY ARE the classic orographic cloud, lenticularis are certainly not the only ones to form orographically. Air rising to pass over raised ground also encourages more common cloud types to appear. Patches of Stratus cloud often cling mistily to hillsides, like souls of long-deceased ramblers, as moist air swirls up the slopes gently and much more gradually than the swift air streams of the lenticularis. Beautiful 'cap clouds' can balance on mountain summits, sometimes appearing like white kippah skullcaps, at others being splayed in a round flat layer, as if the mountain is spinning plates. 'Banner clouds' can unfurl just behind mountain peaks, fluttering white in the wind, as if the mountain has had it with the plate spinning and is surrendering.

Puffy Stratocumulus layers can develop at the top of plateaus,

Roger Colbeck (member 68)

A 'cap cloud' draped over the Mount Blanc massif in the French Alps.

such as the Table Mountain in Cape Town, South Africa, which is often shrouded in what is known locally as the Tablecloth cloud. Its appearance can be explained in terms of the Cape Doctor – the summer wind that tends to blow away the city's pollution as it comes in from the southeast. Having gathered moisture over warm waters on its route towards the city, the air stream forms cloud as it cools by rising over Table Mountain. The Tablecloth cloud is laid out like thick damask across the mountain's surface. But that is not the only explanation for it.

The other involves Jan Van Hunks, an eighteenth-century Dutch pirate, who retired from his busy life terrorising the seas to settle on the slopes of Table Mountain. Van Hunks had a nagging wife, and would often walk up the mountain in order to smoke his pipe in peace.

One day, as he sat there, a stranger approached and asked for some tobacco. The two passed the time puffing away, until the stranger challenged Van Hunks to a smoking contest. He agreed, the prize for the victor being a ship of gold.

Several days passed with the two locked in an intense smoke-off, until Van Hunks finally emerged, coughing and wheezing, as the victor. His pride at his considerable achievement was soon dampened, however, on discovering that the stranger was in fact the Devil. What are the chances of that? Never one to honour his debts, the Devil made the clouds close in around Van Hunks and, with a clap of thunder, spirited him away.

All that was left behind following their contest was a huge cloud of smoke, and this is what became the Tablecloth. In the summer months of November to February when the cloud appears, it's said that Van Hunks and the Devil are at it again with a rematch of their smoke-off.

☁

CLOUDS ARE FOR DREAMERS, and the contemplation of their shapes is a pursuit worthy of any cloudspotter. They are the Rorschach images of the sky – abstract forms onto which we project our imaginations. Time spent considering cloud simulacra that you see is guaranteed to save money on psychoanalysis bills.

Finding shapes in the clouds is practically a full-time occupation for children. Why do so many of us give it up as adults? Any cloudspotter who has become too sensible to see shapes in the

Jean Cassidy (member 250)

Most Altocumulus clouds are not of the UFO-shaped lenticularis species, but in layers of cloudlets, like this stratiformis species.

clouds needs to re-evaluate. They should park their rational minds, and let whimsy take over. I hope they won't just see what they are told to, like the fawning courtier, Polonius, in Shakespeare's *Hamlet*:

> HAMLET: *Do you see that cloud that's almost in shape like a camel?*
> POLONIUS: *By the mass, and it's like a camel, indeed.*
> HAMLET: *Methinks it is like a weasel.*
> POLONIUS: *It is back'd like a weasel.*
> HAMLET: *Or like a whale?*
> POLONIUS: *Very like a whale.*[1]

Perhaps they'll see 'a centaur, or a leopard, or a wolf, or a bull',[2] like the Socrates character in Aristophanes' play, *Clouds*. Perhaps they'll divine 'giants' countenances... great mountains and rocks... after them some monster pulling and dragging other clouds',[3] like Lucretius, the Roman poet, in his philosophical epic *De Rerum Natura* (*On the Nature of Things*).

The Greeks and Romans appear to have been keen enthusiasts of this pastime. Philostratus, the Greek sophist with possibly the best name for making pronouncements on clouds, wrote a romanticised account of the life of a philosopher, called Apollonius of Tyana, who discussed cloud shapes with his sidekick Damis and rather took the magic out of simulacra-seeking. *Cloudshapes of classical antiquity* The two agree that, just as painted images are nothing but make-believe since they consist in fact of nothing but pigments, so the images seen in clouds are make-believe also. Must we then assume, asks Apollonius, that God is an artist, who likes to pass the time painting images in the clouds? Myself, I am rather seduced by the idea, but Apollonius decides otherwise: the shapes in the clouds are produced at random, without the need for any divine intervention. It is nothing more than man's fondness for make-believe that leads him to see faces and animals in them.

The classical preoccupation with cloud simulacra is not something that was ever expressed in their art. In fact, the skies of ancient landscape paintings are, on the whole, entirely devoid of any of our fluffy friends whatsoever. For cloud shapes in art, one has to look to the Renaissance.

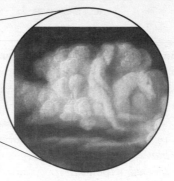

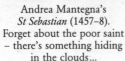

Andrea Mantegna's
St Sebastian (1457–8).
Forget about the poor saint
– there's something hiding
in the clouds...

ONE THE EARLIEST examples is Andrea Mantegna's *St Sebastian*. Cloudspotters who visit Vienna's Kunsthistorisches Museum to see the Renaissance canvas will, of course, ignore the unfortunate arrow-pierced saint in the foreground and look to the sky behind. Mantegna could not be considered one of the great cloud painters, as he had a tendency to get his cloud forms rather muddled. He often painted them with very flat Cumulus mounds at the centre, but with wisps at each end that look more like the high, ice-crystal clouds called Cirrus. Clouds do form into clumps up at the level of Cirrus clouds – they are called Cirrocumulus – but they certainly don't look anything like Mantegna's confused formations.

Behind St Sebastian, however, Mantegna did paint one cloud with a more faithful Cumulus form than the others, and it is here that the keen-eyed cloudspotter will notice something peculiar. Within the cloud's relief, Mantegna depicted the shape of a man on horseback. Why he put it there is something of a mystery – it bears no obvious relevance to the rest of the painting. Perhaps it was no more than a playful reference to the Greek and Roman love of cloud shapes, for the painting contains many classical references, and Mantegna even signed his name in Greek.

He painted *St Sebastian* when he was in his twenties, but it wasn't until Mantegna was an old man of 71 that he once more hid a shape in the clouds. That was in his allegorical *Minerva Expelling the Vices from the Grove of Virtue*, which is in the Louvre, in Paris. This time, it was the profile of a face looming from the Cumulus up at the top of the canvas.

Mantegna's clouds may leave a bit to be desired from a meteorological standpoint, but it should be said that none of the Renaissance artists made them look very realistic. It was the Dutch masters of the Baroque period – most notably, the brilliant Jacob van Ruisdael – who were the first to introduce lifelike clouds in their paintings.

But then, accuracy isn't everything. Sure, Mantegna's cloud hybrids were not a patch on Ruisdael's dramatic, voluminous Stratocumulus and Cumulus. But he clearly had fun hiding shapes in them. No doubt, he enjoyed the thought that many of the viewers would miss his hidden cloud simulacra.

I expect most of them do. Except, that is, for the cloudspotters.

☁

BESIDES THE UFO-SHAPED lenticularis species, there are many other formations of Altocumulus. A cloudspotter can think of the genus as a higher version of the Stratocumulus. Typical Altocumulus clouds – those not formed orographically – are composed of many individual cloudlets spread out in a layer of tessellations. One difference from the lower Stratocumulus cloud is that these cloud elements are often smoother and more regular in size and spacing, as they tend to be out of reach of the chaotic thermals and eddies coming off the Earth's surface. Indeed, an Altocumulus can sometimes be so regular as to resemble a batch of segmented bread rolls – ones which appear to brown with the warm colours of the setting Sun.

There are a couple of easy rules of thumb to distinguish Altocumulus from other cloud types, and the first is more literal than most. This involves holding up your hand to compare the size of the cloud elements against the width of your fingers.

As a lover of clouds, I find it hard to stand in the street sticking my fingers up at them, but doing just that is the easiest way to determine the general height of a layer of cloudlets. Fortunately, it is not a V-sign that is required.

A cloudspotter should hold up three fingers with an extended arm. If the individual elements of the layer are wider than all three fingers, the cloud is probably of the lower Stratocumulus genus. If they are smaller than the width of one finger, then it is more likely to be a high layer of cloudlets, called a Cirrocumulus. It is most likely to be an Altocumulus layer when the size of the cloudlets is somewhere between the two – smaller than three fingers and larger than one.

However, the giving-the-cloud-the-fingers rule doesn't work if cloudspotters are looking at clouds off in the distance. Their outstretched arm needs to be angled above 30° from the horizontal for it to apply. Cloudspotters should resist measuring cloudlets against all five digits, as they will be in danger of performing the Nazi salute.

The second rule of thumb for identifying Altocumulus has to do with the shading of its cloudlets. When the sky above is clear and the Sun shines directly on to the cloud, Altocumulus will have noticeable shading on the sides away from the Sun, though this will not be particularly heavy. With the lower Stratocumulus, the shaded parts are often quite dark, while the tiny cloudlets of the high Cirrocumulus show no shading at all. In other words, small cloudlets with visible shading mean the cloud is an Altocumulus.

In the air below Altocumulus elements, there commonly form supplementary features, known as virga. These are fibrous trails of precipitation that appear to hang below the cloudlets, making them look rather like jellyfish. They occur when rain or snow is falling from the cloud, but has evaporated away before reaching the ground. Their presence can help to distinguish Altocumulus from a collection of small Cumulus clouds (Cumulus humilis), which never produce precipitation. Noticing the virga tendrils, any cloudspotter can confidently turn to a passing stranger and say, 'Those jellyfish are Altocumulus, you know, not a gathering of high Cumulus clouds that you might think they are.'

The form and arrangement of the Altocumulus cloudlets can be wonderfully varied. In fact, this genus exhibits even more species

Michael Rubin (member 329)

Individual Altocumulus cloudlets exhibiting 'virga' – the trails of evaporating precipitation that make them look like jellyfish.

Alex Sinclair (member 3267)

Altocumulus stratiformis translucidus. Like most of the Altocumulus species, the cloudlets appear between the width of one and three fingers, held at arm's length.

and varieties than you find with the Cher-like costume changing of the lower Stratocumulus. It exhibits the same seven official varieties as the Stratocumulus – opacus, translucidus, perlucidus, duplicatus, undulatus, radiatus and lacunosus – but it has one more official species than its lower cousin.

Both exhibit the UFO-shaped lenticularis species, the commonly occurring stratiformis, where the layer extends over large areas of the sky, and the castellanus, with its castle-like crenellations at the top of the cloudlets. But the Altocumulus can also be described as floccus. This is the species in which the tops of cloudlets have Cumulus-like mounds to them, rather than the usual smoother surfaces. Altocumulus floccus are often accompanied by virga falling from their ragged bases.

☁

THERE IS AN ART to seeing shapes in clouds. It comes easily to some but can present quite a challenge for others. Children tend to do it effortlessly, though it was never that way for me.

I remember the day when, aged four or five, we 'did clouds' at school. Marched in crocodile formation to the grass behind the school, we lay down on our backs and looked up at the cloud formations. After conducting an involved analysis of the types present ('fluffy ones', 'little ones', 'a bit bigger ones', 'wispy ones') we were invited to see if we could find any shapes in them.

Our teacher – I seem to remember that she was called Mrs McCloud, but I might be making that bit up – read a story about someone seeing shapes in the clouds. And then my classmates started calling out what they could see. Sean saw a dragon. Jessie was overjoyed to have spotted a flower. I couldn't see anything – just lots of fluffy clouds.

I stared and stared but, for the life of me, couldn't find any shapes at all. There were superheroes being spotted over here, strange fish faces over there, and it was beginning to stress me out. I asked the little girl next to me where the head of the mermaid was. She pointed her finger and said it was next to the puffy bit. The more I searched for shapes, the more frustrated I became. Why couldn't I see anything? I just wanted to get a rake and sweep all the clouds away, as I'd seen Ermintrude do in *The Magic Roundabout*.

After the shape finding was over, we were herded back into the classroom, where we stuck tufts of cotton wool on to blue paper. They resembled Cumulus clouds at the humilis and mediocris stages of vertical growth. Mrs McCloud invited us to add little plastic beads to make our clouds rain.

Anyone who has been paying attention this far will know what a crime it was to encourage this – it is, of course, only in their congestus and Cumulonimbus stages that convection clouds typically produce precipitation. In spite of this meteorological *faux pas*, the cotton wool part of the day's activity was thankfully one I managed without mishap. The whole shapes débâcle stuck with me, however. I couldn't help worrying that there was something wrong with me: I couldn't see shapes in the clouds when everyone else in the world could.

☁

MOST CLOUD FORMATIONS look their most dramatic at sunrise and sunset. But the mid-level Altocumulus stand out from the crowd as being more beautiful than any other in the low-angled sunlight of these times of day. The elements of the cloud layer are clumpy enough for the low Sun's rays to cast them into dramatic relief – some parts ablaze with reds, oranges and pinks and others shaded in deep indigos. Moreover, its mid-level altitude means it is high enough for the Sun to shine on to it from below, without being so far away that the elements are difficult to make out.

Cloud specialists are often a bit sniffy about sunrise and sunset photographs. They are, I suppose, just too obvious: any Tom, Dick or Harry will gasp at a sunset, but only a true connoisseur will appreciate the unique beauty of a pileus accessory cloud delicately perched atop the billowing mounds of a Cumulus congestus. Whatever encourages people to look up is a good thing, as far as I'm concerned. Oscar Wilde, however, presented a particularly snooty view of sunsets in his 1889 dialogue *The Decay of Lying: An Observation*:

Carolyn Torella (member 1634)

Altocumulus, or a dove.

Michael Poole (member 1594)

Cumulus fractus, or Bob Marley.

Nobody of any real culture… ever talks nowadays about the beauty
of a sunset. Sunsets are quite old-fashioned. They belong to the time
when Turner was the last note in art. To admire them is a distinct sign
of provincialism of temperament. Upon the other hand they go on.
Yesterday evening Mrs. Arundel insisted on my going to the window,
and looking at the glorious sky, as she called it. Of course I had to look
at it… And what was it? It was simply a very second-rate Turner,
a Turner of a bad period, with all the painter's worst faults exaggerated
and over-emphasised.[4]

☁

CLOUDSPOTTERS MUST SURRENDER themselves to the
gentle shifts of the clouds' formations. If they cannot identify a
particular cloud form, then so be it – they should just relax and
watch it develop. Before long, a familiar formation will appear.
Now that I look back on my frustration at being unable to divine
cloud shapes as a child, I realise I was going about it the wrong way.

I was straining to keep up with Sean and Jessie and my other
classmates. The more I searched for the simulacra, the more elusive
they became. The more I struggled to make out a nose or a trunk,
the more formless and abstract the clouds seemed.

This is not how to look at clouds. Cloudspotters won't find
shapes in them by force of will, nor by searching with half a mind
on what the person beside them sees. The best way to find shapes
is to look up, empty your mind, and let them find you.

I imagine that many of those classmates no longer look for
cloud shapes. No doubt Jessie is too busy with the afternoon rush
to collect her own children from school, Sean too preoccupied
with keeping an eye on the budget for his family's house extension.
I imagine they have neither the time nor the patience any longer to
find space in their days for cloud shapes to appear. The good news
is that I finally seem to be getting the hang of it.

This has come about by not caring whether I see anything in
the clouds – by delighting in the very formlessness of the drifting
skies. Now I know better than to try and will the shapes into
existence. It seems pertinent that Aristotle used the metaphor of

Michael Rubin (member 329)

Cumulus mediocris, or two cats dancing the salsa.

cloud shapes when he attempted to explain what happens when we dream. Dreams, he noted, occur only when the sensory hubbub of consciousness is removed:

> *They are within the soul potentially, but actualise themselves only when the impediment to their doing so has been relaxed; and according as they are thus set free… they possess verisimilitude after the manner of cloud-shapes, which in their rapid metamorphoses one compares now to human beings and a moment afterwards to centaurs.*[5]

It seems that cloud shapes also only appear when the impediment to *them* doing so has been relaxed. You can't will dreams to happen, nor can you force cloud shapes to appear.

☁

Steve Flitton (member 342)

A dissipating Cumulus tower, or Thor, the Norse god of thunder.

IS THE ABOVE A PHOTOGRAPH of the decaying towers of a Cumulus congestus that have just precipitated away much of their moisture? Or is it the Norse god Thor, wielding his thunder hammer with his daughter Thrud under his arm? Of course, it is both. There are two ways of looking at clouds – two ways of seeing – both of which a cloudspotter should cultivate, for each is as valuable as the other. Noticing that a cloud is in the shape of a jellyfish is as worthy an activity as identifying it to be an Altocumulus floccus exhibiting virga. One is, after all, no more than a Latin version of the other.

SIX

ALTOSTRATUS

*The mid-level layers,
known as 'the boring clouds'*

The Altostratus is a mid-level layer cloud and, like its lower equivalent, the Stratus, it is not known for its beauty. It is typically a featureless layer, often stretching across the whole sky and forming at heights of between 6,500 and 23,000ft.

The Altostratus can be a challenge to identify, since it shares some characteristics of the Stratus, and some of the higher layer cloud called the Cirrostratus. It is a middle-of-the-road cloud, which comprises neither solely liquid water particles nor ice particles but often contains a combination of the two, depending on the air temperatures. It is a smooth blanket of cloud that can vary greatly in depth – appearing as a bright, blue-white layer when it is thin enough for the sky to partially show through and as a darker grey one when it is thick and opaque. Nor can the Altostratus be conclusively described as either a dry cloud or a precipitating one. Although it is often just light grey overcast sky that causes no precipitation at all, in its thicker incarnations it can produce a continuous, steady fall of light rain, snow or ice pellets. Given that it's such an in-between cloud, you can see why identifying an Altostratus can seem a rather grey area.

How, then, is a poor cloudspotter to know if he is looking up at one? It is easier if the Altostratus is of the translucidus variety, meaning that it is thin enough for the position of the Sun or Moon to be distinguishable through it.

ALTOSTRATUS CLOUDS

Altostratus are mid-level layers of grey cloud, which are either featureless or fibrous in appearance, and typically extend over an area of several thousand square miles. Usually composed of both water droplets and ice crystals, they are often thin enough in parts to reveal the position of the Sun, which appears as if through ground glass. Altostratus can cause a white or (when very thin) coloured 'corona' (disc of light) around the Sun or Moon.

TYPICAL ALTITUDES*:
6,500–23,000ft
WHERE THEY FORM:
Worldwide. More common in the middle latitudes.
PRECIPITATION (REACHING GROUND): Usually not, but occasionally light rain or snow.

ALTOSTRATUS SPECIES:
There are no species, as the cloud's appearance is so uniform.

Altostratus translucidus

Altostratus radiatus

ALTOSTRATUS VARIETIES:
OPACUS: When the cloud layer is generally thick enough to mask the position of the Sun or Moon.
TRANSLUCIDUS: When it is generally thin enough to show the position of the Sun or Moon.
DUPLICATUS: When there is more than one layer at different altitudes, these often being partly merged. This is generally only visible when, by the light of a low Sun, the higher layer is lit and the lower is in shadow, or when shearing winds cause the striations of the layers to differ.
UNDULATUS: When the layer shows largely parallel undulations.
RADIATUS: When lengthy undulations appear to converge towards the horizon.

NOT TO BE CONFUSED WITH...
CIRROSTRATUS: which is a higher layer of ice crystals that looks like a thin, milky veil across the sky, and often thickens and lowers to develop into Altostratus. The Altostratus will tend to be more opaque, making the sunlight too diffuse for objects to cast shadows, as they do below Cirrostratus. While coloured or white discs of light, called coronae, can appear around the Sun/Moon through Altostratus, this cloud will not cause the 'halo phenomena' of the Cirrostratus.
NIMBOSTRATUS: which is a thick, dark layer of precipitating cloud that often develops out of an Altostratus. Generally darker, it produces considerably heavier rain or snow.

* These approximate altitudes (above the surface) are for mid-latitude regions.

The most likely error is to mistake it for either the lower Stratus cloud or the higher Cirrostratus. Compared with the Stratus translucidus, the Sun seen through an Altostratus layer appears as if through ground glass: it has a more diffuse, blurred look to it, due to the cloud's combination of liquid droplets and solid ice particles. When the Sun or Moon is shining through a Cirrostratus, which is often composed of prism-shaped ice crystals, it can cause optical effects such as rings of light or 'halos'. The Altostratus hardly ever produces a halo, so the mere presence of these is enough to distinguish the two.

A thin Altostratus does, however, show an optical effect of its own around the Sun or Moon. This is called a 'corona', and appears as a solid disc of light – very different from the ring of the Cirrostratus's halo. The disc of the corona is a bright, bluish colour, which sometimes merges to a yellowy white towards its circumference, with a brownish outer edge. Further, more faint rings of colour can occasionally be seen outside the disc when the Altostratus's droplets have a particularly uniform size.

The presence of a corona is not much help in distinguishing the two clouds, however, for Cirrostratus can exhibit them as well as Altostratus. Nevertheless, it is fair to say that if a halo is present around the Sun, then the cloud is a Cirrostratus, not an Altostratus.

The other way of distinguishing an Altostratus from a Cirrostratus is by looking not up but down. The sunlight through a layer of the higher Cirrostratus is always sharp enough for cloudspotters to see their shadow on the ground, whereas an Altostratus will tend to leave them, like Peter Pan, bereft of one.

But what if the position of the Sun or Moon is not visible through the Altostratus? What is the cloudspotter to do when the cloud layer has grown thick enough to be of the opacus, rather than the translucidus, variety, so that it doesn't show the position of the Sun or Moon? Well, firstly one can be sure that it is not the higher Cirrostratus cloud, since that always lets enough light through to see them. More tricky is distinguishing it from the lower, equally thick Stratus opacus.

The Altostratus is often even more featureless than the Stratus. If a cloudspotter can see any texture to the base of the cloud deck,

It is not hard to see why some call Altostratus the boring cloud.

it is more likely to be the lower-level Stratus. From the ground, however, there is not a lot in it. The only way to be sure is to determine the altitude of the opaque layer of cloud – below 6,500ft and it's Stratus, between 6,500 and 23,000ft and it's Altostratus. But since cloudspotters don't usually have a plane to nip up in, distinguishing between thick Altostratus and Stratus clouds is the greyest part of an already grey area.

With such a paucity of visual characteristics, the Altostratus genus is not divided up into species. There are, however, five official varieties: the translucidus and opacus, which depend on whether you can see the position of the Sun or Moon; duplicatus, when there is more than one distinct layer; undulatus, when it has gentle undulations just distinguishable on its underside; and radiatus, when these stretch far enough for them to appear to converge off towards the horizon.

This mundane cloud, it has to be said, has not been blessed with much to recommend it. Meteorologists joke that it is the boring cloud, and I can see what they mean.

☁

BUT EVERY CLOUD has its day. Or at least its time of day. And for the Altostratus, like many, this is at sunrise or sunset. Any cloud – even the most drab – becomes more beautiful in the light at dawn and day's end.

No one knew this better than the American naturalist, Henry David Thoreau. He was a big fan of the clouds in general, claiming that 'The most beautiful thing in nature is the sun reflected from a tearful cloud.'[1] His journals are peppered with eulogies to the sky – none more enthusiastic than those after a spectacular sunset. Thoreau knew that the sunset's beauty is thanks to the clouds:

> *Those small clouds, the rearmost guard of the day, which were wholly dark, are again lit up for a moment with a dull-yellowish glow and again darken; and now the evening redness deepens till all the west or northwest horizon is red; as if the sky were rubbed there with some rich Indian pigment, a permanent dye; as if the Artist of the world had mixed his red paints on the edge of the inverted saucer of the sky...*
> *It is like the stain of some berries crushed along the edge of the sky.*[2]

For fleeting moments when the Sun is low on the horizon, even the boring Altostratus dons her fancy clothes to paint the sky red with ruby hues. It doesn't last long, of course, but what a difference it makes! Suddenly, this unremarkable grey sheet is transformed into a beautiful sea of delicate salmons, pinks and mauves.

The low angle of the Sun also accentuates any contours to the Altostratus's base, bringing much-needed relief. 'The inhabitants of earth behold commonly but the dark and shadowy underside of heaven's pavement;' wrote Thoreau, 'it is only when seen at a favorable angle in the horizon, morning and evening, that some faint streaks of the rich lining of the clouds are revealed.'[3]

Of course, the Sun only has a chance to light the underneath of a layer such as the Altostratus if the skies are clear off in the direction where it is rising or setting. This accounts for the age-old saying 'Red sky at night, shepherd's delight. Red sky in the morning, shepherd's warning.' Weather systems in temperate regions tend to move from west to east*, and so a red sky at night means that, though there are clouds above you, it is fairly cloud-free to the west (where the sun is setting), which allows the ruby light to flood up below them. Clear skies to the west mean there is a fair chance of clear weather approaching. Conversely, a red sky in the morning means that the clouds above are lit with the Sun's morning hues through clear skies to the east so there is a fair chance the clearer weather has passed and more cloud is on its way.

The observation has been around for millennia. According to the Bible, Jesus mentioned it when he rebuked the Pharisees and the Sadducees, who had asked him for a sign from heaven:

> *He answered and said unto them, When it is evening, ye say, It will be fair weather: for the sky is red.*
> *And in the morning, It will be foul weather to day: for the sky is red and lowring. O ye hypocrites, ye can discern the face of the sky; but can ye not discern the signs of the times?*[4]

I can't help wondering if the same small clouds, that Thoreau called the 'rearmost guard of the day', gave rise to the name of cloudberries. These are related to raspberries, but have fewer segments, making them similar in shape to the cloudlets of the Altostratus's mid-level relation, the Altocumulus. As they ripen, the berries change colour from a rich, fiery red to saffron and gold. It is the same shift of hues that an Altocumulus cloud exhibits in the brightening light of the sunrise.

Cloudberries only grow on boggy land, often in remote northern climates, which makes them something of a delicacy. The vast wilderness of the tundra regions of Arctic Russia is just the sort of place in which they thrive. Through the chilly 24-hour days of the summer months, they ripen slowly – gradually developing their unique, intense flavour – and stand out like tiny sunset Altocumulus against the dark mosses and shrubs on the wetlands of the tundra.

....

* This is due to the combined effect of the poleward decline of temperature and the Earth's rotation on the movement of air around the globe.

Mike Davies (member 1633)

It is in the warm colours of dawn and day's end that the Altostratus cloud shines, as demonstrated in this black and white photograph.

☁

ON A CHILLY EVENING towards the end of September, however, it is a flat blanket of Altostratus that is draped across the skies above the Pechora Delta, where Siberia meets the Barents Sea. The cloud has its party clothes on, for with the approach of winter, the days are shortening and the Sun has finally begun to dip below the horizon.

There are few inhabitants of this remote region to enjoy the cloudberries, but they certainly don't go to waste. Olya, like many of the thousands of visitors who come here each year, is stocking up on them. She has spent the last fifteen summers in the delta and, below the vermilion-coloured Altostratus, she eats cloudberries as if her life depends on it. Which, I suppose, it does.

Olya is a swan – a Bewick's swan, to be precise – and she's been fattening herself up on cloudberries, along with a host of the tundra's abundant shrubs and mosses, with one purpose in mind. Tonight, she is setting off with her mate and their two young cygnets on the start of a 2,000-mile migration to the Severn Estuary in Gloucestershire. With various rest stops along the way, the journey should take them a couple of weeks. It will be a gruelling trip, which makes it rather touching that she's coming all this way to demonstrate a principle of optics.

Keith Epps (member 868)

Undulatus is one of the more interesting varieties of Altostratus, even without the colours of a low Sun.

You see, Olya is going to show us why the light cast from a low Sun adorns clouds like the Altostratus in such glorious fiery colours. She is coming all this way to reveal why sunlight passing at a low angle through the atmosphere changes its hue.

OK, I know ornithologists will argue that Bewick's swans like Olya have other reasons for leaving the Arctic tundra in September. The temperature drop at this time of year heralds the start of the bitter Siberian winter, when the whole region begins to freeze over and the cloudberries, like all the other vegetation, become a distant, bird-brained memory. The Bewick's swans are creatures of habit, they'll tell us, who teach their young cygnets the route to their favoured winter feeding grounds in the wetlands of northwest Europe. Having found a suitable spot, they tend to return each year – even if it is two thousand miles away.

That may all be so, but Olya is different. She is on a mission. Her optics demonstration will take place when she and her family reach their destination amongst the reeds of the Severn Estuary – a region of the world where cloudberries are nowhere to be found.

☁

LIKE THE OTHER SWANS vacating the tundra, Olya and her mate have chosen the evening of their departure carefully. Throughout the day, a northeast wind has been blowing across from the Kara Sea, past the archipelago of Novaya Zemlya, and it promises to assist their westerly flight along the coast. Moreover, the Altostratus cloud this evening is high enough and thin enough to afford them good visibility as they fly through the night. They will make many stops at lakes and marshes to rest and refuel over the coming weeks. Progressing southwest over the White Sea and across Karelia to the Gulf of Finland, their route will hug the Baltic coast and they will finally cross the North Sea from the Netherlands to reach the coast of Britain.

Whilst swans – even Bewick's, which are smaller than most – are ungainly in take-off, they are more graceful in landing. This is just as well, for it means that, on arrival in Gloucestershire, Olya can give her demonstration in the style one would expect from a swan, even one that's weary from such a mammoth journey and considerably thinner than when she left.

Her mate, being the stronger flier, is at the front of the family formation as they finally reach the reedy marshes of the Severn Estuary, arriving with the dawn on a mid-October morning. Olya's mate doesn't care for physics and so he lands out in the water, well away from the reeds. But this is Olya's moment, and she chooses her landing spot well.

Splaying her webbed feet out before her, she swooshes down elegantly on to the surface of the still estuary near to where a patch of reeds is growing. A wide wave billows from her and passes into the reeds, which act as small obstacles in its progress along the water's surface.

In the same way, the microscopic particles that make up our atmosphere – all the molecules of oxygen and nitrogen and particles of dust, salt and soot – act as obstacles to the waves of sunlight that come to Earth from the Sun.

The demonstration, which Olya has travelled these two thousand miles to perform, has two parts to it, and she has just completed the first. The wave that came off her as she landed this fresh October morning was considerably wider from crest to crest than the little reedy obstacles in its path. It therefore passed largely unimpeded through the reeds, to emerge the other side.

Of all the visible light waves that come to Earth from the Sun, those that look red to us are the ones that have the longest distance between their peaks – the longest wavelengths. Just like the big wave that came off Olya as she landed, these longer wavelengths of light tend to pass through the atmosphere without being scattered much by all its tiny molecules and particles.

Now it is time for Olya to progress to the second part. She's unsettled and hungry, but she is going to do it anyway. Having quickly checked that her two cygnets have landed safely, she now paddles, ever so gently, right past the same patch of bulrushes, sending short ripples across the water from her chest.

Unlike the wide waves, when these ripples reach the reeds they have wavelengths (from crest to crest) that are about the same size as the thickness of the reed stems. 'See how they are scattered by

Name that Cloud!

Cloudspotters should undertake this test after reading the guide from cover to cover and notching up many hours of carefree cloud gazing. It will guarantee a nebulous sense of achievement. Answers given at the end.

Angelo Storari (member 2378)

1. Name this cloud. (I don't mean give it a name – like 'Philip' – merely, identify its genus and species.) And – for a bonus point – what is the name for the shafts of light emanating from behind it?

Alex Raistrick (member 1712)

2. This sky contains a number of different cloud types, but what species is hovering down towards the bottom and bottom right of the image?

Richard Atkinson (member 11)

3. Whilst they don't have an officially-recognised name, the Cumulus clouds that form in the hot, moist fumes of power stations do have a nickname. Can you guess what it is?

4. The main cloud type here is Cirrus, and it has the appearance of a cloud in danger of a nervous breakdown. With such chaotic orientations, what variety of Cirrus is it: radiatus, intortus, vertebratus or duplicatus?

5. What is it that's so pleasing about this layer of Stratocumulus?

6. Some might think that this looks like a lenticularis species, but it isn't. It is a formation that can result from the winding streams of high winds that circle the globe in the mid-latitudes. What are the winds called?

7. There are some 'mamma' (at the top of the picture) appearing in this cloud, which is probably the anvil of a Cumulonimbus. But what is going on with that weird light effect?

Both images: Peg Zenko (member 1527)

8. These are two different types of 'halo phenomena', which sometimes occur as sunlight passes through high, ice crystal clouds (in these examples Cirrostratus). Which halo phenomena are shown here?

Dr Peter and Mrs Gill Smith (members 1522)

9. Nacreous, or 'mother-of-pearl' clouds, form above the troposphere. At what range of altitudes are they typically observed and, if not the troposphere, then in which region of the atmosphere?

Nick Lightbody (member 95)

10. Which rather dull cloud here looks uncharacteristically resplendent in the warm rays of the rising Sun?

11. What cloud is this?

....

HAVE YOU PASSED THE
CLOUDSPOTTING DIPLOMA?
HERE ARE THE ANSWERS...

1. It is a Cumulus cloud and, since it is wider than it is tall, it is of the humilis species. The shafts of light emanating from behind it are called 'crepuscular rays'.

2. It is the lenticularis species – in this case, most probably an Altocumulus lenticularis – which can form when moist airstreams take on a wave-like motion in the lee of hills or mountains.

3. 'Fumulus'.

4. Intortus.

5. It is pleasing for whatever reason you find it to be.

6. Jet streams. These are known as 'jet-stream Cirrus'.

7. The light effect is nothing more than the shadow of the Cumulus congestus cloud off towards the horizon.

8. Top: A circumzenithal arc (CZA).
   Bottom: A parhelium, also known as a sundog – though in this case, since the light is being cast from a full moon, it is strictly called a 'moondog'.

9. 10–15 miles up – in the part of the atmosphere called the stratosphere.

10. Altostratus.

11. It is an 'Abominable Snowman, who is upset that his pet seahorse is ignoring him' cloud.

THE COLOURS OF A LOW SUN

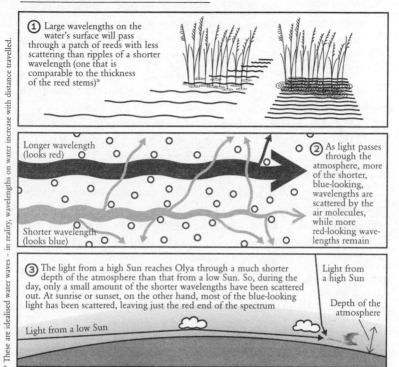

① Large wavelengths on the water's surface will pass through a patch of reeds with less scattering than ripples of a shorter wavelength (one that is comparable to the thickness of the reed stems)*

Longer wavelength (looks red)

Shorter wavelength (looks blue)

② As light passes through the atmosphere, more of the shorter, blue-looking, wavelengths are scattered by the air molecules, while more red-looking wavelengths remain

③ The light from a high Sun reaches Olya through a much shorter depth of the atmosphere than that from a low Sun. So, during the day, only a small amount of the shorter wavelengths have been scattered out. At sunrise or sunset, on the other hand, most of the blue-looking light has been scattered, leaving just the red end of the spectrum

Light from a high Sun

Depth of the atmosphere

Light from a low Sun

* These are idealised water waves – in reality, wavelengths on water increase with distance travelled.

Olya demonstrates why clouds look red and orange at sunrise and sunset.

the reeds?' she would say if she could, for as they enter the patch of reeds, they are dispersed.

The part of visible sunlight that looks blue and violet to us has wavelengths comparable in size to the molecules and particles that make the atmosphere. And, like Olya's ripples, those waves are scattered more as they pass through the atmosphere.

As Olya paddles off for some food and rest, introducing her cygnets to their new winter home, cloudspotters might well be wondering how her demonstration explains the warm colours of clouds like Cirrostratus at sunset and sunrise.

When the Sun is high in the sky, its light reflects off the clouds and arrives at ground level having passed down through the depth of the atmosphere. Pretty much the full range of visible light reaches us, for the amount of atmosphere it passes through is not enough to cause too great a proportion of the short wavelengths – the ones that look blue and violet – to have been scattered away. The full spectrum of visible light appears white to us, and so therefore do the clouds that reflect it.

When the Sun is low on the horizon, the light reflected off the clouds only reaches us once it has passed through a long, tangential slice of the atmosphere. Indeed, since the atmosphere tends to bend the light around the curve of the Earth, rays from near the horizon can pass through as much as forty times more atmosphere than those coming straight down from a high Sun.

At this angle, the light reflects off the clouds and reaches the cloudspotter only after most of the short (blue- and violet-looking) wavelengths have been scattered away by the molecules and particles in the atmosphere. The longer, red-looking wavelengths make it through largely unimpeded.

The same principle explains the blue of the clear sky during the day. The only visible light that reaches us from directions other than that of the Sun has been scattered our way by the molecules and particles of the atmosphere. These are primarily the shorter wavelengths that appear blue and violet. (Our eyes are less sensitive to the shorter, violet-looking wavelengths, so the blue colour is dominant.) It also explains why sunsets are deeper and redder after volcanic eruptions. All the additional particles thrown up into the atmosphere scatter even more of the short and middle wavelengths, leaving a narrower spectrum of red-looking wavelengths.

Red clouds were considered particularly favourable signs in Ancient China. Apparently, one emanated from Lao Tzu, the philosopher credited with founding Taoism. Red and yellow were the colours of 'cosmic differentiation', and coloured clouds would come down upon the mounds on which acceptable sacrifices had been offered. In fact, Huang-ti, the mythical Yellow Emperor who supposedly ruled in the third millennium BC, was considered to 'govern all things, thanks to the clouds'.

No less crucially, the colours of the clouds can also give cloudspotters an indication of their relative heights in the atmosphere. The geometry is such that when the Sun is just above the horizon, light reflected off the low clouds will have passed through more of the atmosphere, and so appear redder, than that from high ones. A low Sun therefore colour codes the cloud altitudes: the highest ones being bright white, the mid-level golden and the lowest red. When the Sun is just below the horizon, the lower clouds become dark, in the shadow cast by the Earth.

☁

THOREAU WOULD have had no time for scientific explanations for the colours of the sunset. He saw no need for recourse to dry science to explain the changing colours of the evening. Regardless of their cause, the only important thing was to appreciate them:

> *I witness a beauty in the form of coloring of the clouds which addresses itself to my imagination, for which you account scientifically to my understanding, but do not so account to my imagination… I, standing twenty miles off, see a crimson cloud in the horizon. You tell me it is a mass of vapor which absorbs all other rays and reflects the red, but that is nothing to the purpose, for this red vision excites me, stirs my blood, makes my thoughts flow… What sort of science is that which enriches the understanding, but robs the imagination?*[5]

Thoreau was echoing the sentiments of the poet, John Keats, who hated Isaac Newton for explaining the rainbow with dispassionate reference to wavelengths of light passing through water droplets. For Keats, there was no soul in the 'cold philosophy' of Newton's explanations:

> *Do not all charms fly*
> *At the mere touch of cold philosophy?*
> …
> *Philosophy will clip an Angel's wings,*
> *Conquer all mysteries by rule and line,*
> *Empty the haunted air, and gnomèd mine –*
> *Unweave a rainbow, as it erewhile made*
> *The tender-person'd Lamia melt into the shade.*[6]

Tim Salter (member 1621)

Altostratus sometimes exhibits supplementary features called 'mamma'. These udder-like pouches of cloud add some much-needed relief to the layer's flat grey.

I can see what Thoreau and Keats mean. But they do sound a bit like the arty kids in class taunting the science nerds. Having opted for all sciences at secondary school, I have painful memories of bullying classmates goading me for emptying the haunted air, and gnomèd mine. OK, they might not have quite put it in those words, but the sentiment was the same.

Cloudspotters will float above these petty divides between science and art – float above them like our fluffy friends. For us, there is no contradiction in regarding the clouds in ways that *both* stir our blood by exciting our imagination *and* enrich our understanding with 'cold philosophy'.

When it comes to the latter, I imagine Olya would want to point something out to Thoreau. As she was at pains to show, a cloud does not turn red in the low Sun because it is 'a mass of vapor which absorbs all other rays and reflects the red'.

Had he been paying any attention, Thoreau would have understood that clouds reflect all the visible wavelengths more or less equally. The colour comes from the way the low-angled sunlight has its shorter wavelengths scattered away on its long, tangential journey through the atmosphere. That's why only the red-looking light reaches us.

Bloody hell, she might as well have not bothered with the whole migration thing.

☁

IT SO HAPPENS THAT Altostratus usually forms as a result of the thickening and lowering of the higher layer cloud called Cirrostratus. And, more often than not, the descent of the cloud layer doesn't stop at the Altostratus level.

Accompanied by a steady fall of light precipitation, the Altostratus can continue to thicken, darkening as its nebulous base lowers. With the increased depth to the cloud layer, the precipitation begins to fall more heavily. Soon, the cloud no longer occupies just the mid levels of the troposphere, but extends from there down to just a few thousand feet. The Altostratus has now changed into a Nimbostratus – a cloud that revels in incessant rain.

It doesn't always progress like this, of course. Sometimes the Altostratus develops and just hangs there, wearing the overcast grey of the boring cloud – not sure whether to stay or go, uncertain whether to rain or not.

However dull some individuals may be, it is most generous to remember them for the times when they shine. Cloudspotters should remember the Altostratus for its fleetingly extrovert moments at the start and close of the day. If they do so, this cloud will teach them a useful lesson – one that was well articulated by John Ruskin: 'Let every dawn of morning be to you as the beginning of life, and every setting sun be to you as its close.'

SEVEN

NIMBOSTRATUS

*The thick, grey blankets that rain
and rain and rain*

The Latin for a rain cloud is 'nimbus'. The word is used in the name of the Nimbostratus because the cloud is, by definition, precipitation-bearing. It is also dark, thick and ragged in appearance.

Water falls from many cloud types, but it is only when it reaches the ground (as opposed to evaporating on the way down) that the cloud is officially defined as a precipitating one, and given the name 'praecipitatio'. The term is not needed, however, for the Nimbostratus since it goes without saying that rain, snow or ice pellets, if not cats and dogs, will be reaching the ground below this wet layer cloud.

The word nimbus is also, of course, used in the name of the Cumulonimbus thundercloud. But this is where the similarity ends, for though they are both precipitating clouds they express themselves in very different ways. Cumulonimbus shed their water in fierce storms and are done with it in a matter of minutes, while the Nimbostratus will tend to release its load steadily, over many hours. It is, you see, a rather slow and ponderous cloud.

Its shape is also very different from that of the thundercloud. The Nimbostratus doesn't have the impressive, tall tower so characteristic of the Cumulonimbus. It is a less-than-spectacular layer, which can spread over thousands of square miles. But still waters run deep and, were the two clouds to have an argument that

HOW TO SPOT
NIMBOSTRATUS CLOUDS

Nimbostratus are thick, grey, featureless layers of cloud that cause prolonged, continuous, often heavy, rain, snow or ice pellets. They tend to have very diffuse bases, as a result of all the falling precipitation. Nimbostratus are the deepest of all the layer clouds – sometimes extending from 2,000ft up to around 18,000ft – and generally extend over many thousand square miles. As with other precipitating clouds, the falling precipitation can cause Stratus fractus to form in the air below Nimbostratus clouds. These are known as 'pannus' and appear as shreds of cloud, looking darker than the underside of the Nimbostratus. When these join together, they tend to lower the bases of Nimbostratus clouds even further. They are invariably thick enough to completely hide the Sun or Moon.

TYPICAL ALTITUDES*:
2,000–18,000ft
WHERE THEY FORM:
Worldwide. More common in middle latitudes.
PRECIPITATION (REACHING GROUND): Causes moderate to heavy rain or snow (steady and prolonged).

NIMBOSTRATUS SPECIES:
There are no species, as the cloud's appearance is so uniform.

NIMBOSTRATUS VARIETIES:
There are no varieties, as the cloud's appearance is so uniform.

NOT TO BE CONFUSED WITH...
ALTOSTRATUS: which is a thinner – though also indistinct – layer of cloud. Nimbostratus is always darker than it and, by definition, produces precipitation. Altostratus only does sometimes, and this will generally be light. Whilst the position of the Sun can generally be determined through at least part of a layer of Altostratus, it will never be so through a Nimbostratus.
CUMULONIMBUS: which, when observed from directly below, can also appear as a very dark layer, covering the whole sky. The precipitation falling from a Nimbostratus will not generally be as heavy and will be more prolonged and continuous, compared with the sudden showers of the Cumulonimbus. Nor will the Nimbostratus produce its hail, thunder or lightning.

Nimbostratus – never a pretty sight.

* These approximate altitudes (above the surface) are for mid-latitude regions.

David Foster (member 1157)

ended in a scrap, it is by no means clear that the Cumulonimbus would emerge the victor.

Sure, the puffed-up convection cloud may have all the crowd-pleasing moves – it would employ fancy footwork worthy of Muhammad Ali and duck and dive like a sprightly Prince Naseem – but it's the Nimbostratus that would get my bet. Though it doesn't have the vertical reach of the King of Clouds, it more than makes up for this with its enormous coverage and steady, relentless endurance. In spite of the thundercloud's enormous punch, I feel this solid performer would hang in there and win out in the end.

The Nimbostratus can consist of any combination of water droplets, raindrops, ice crystals and snowflakes, depending on air temperatures. Clouds have to be tall to produce much precipitation – as meteorologists describe it, they have to be 'deep' – and the Nimbostratus tends to extend through more than one of the low, middle and high *étages* into which most clouds are categorised. Though it never grows as tall as the Cumulonimbus, it can occasionally reach from just 1,000ft above the ground right up to the higher reaches of the middle level, around 20,000ft.

Who would win?

For this reason, the cloud level in which to group the Nimbostratus is a rather moot point. It does, however, invariably inhabit *at least* the middle level, so I'm sure it will be happy enough to be discussed in the company of the other mid-level clouds.

☁

IF IT IS SUCH A HEAVYWEIGHT, and able to stand its ground against the King of Clouds, then why is the Nimbostratus as good as unknown outside of meteorological circles? The Cumulonimbus is, of course, a household name, but mention the Nimbostratus to the average person and you will invariably draw a blank.

No doubt this is because it's not a very interesting cloud to look at. Standing below a developing Nimbostratus, a cloudspotter will see no more than the ominous darkening and lowering of its featureless base. As the poet John Milton put it, '...O'erspread / Heaven's cheerful face, the louring element / Scowls o'er the darkened lantskip snow or shower'.[1]

As the cloud layer thickens, metamorphosing from a higher, thinner Altostratus, it blocks out more and more of the light so that the overcast sky darkens. By the time it has become the precipitating Nimbostratus, it is so thick that the position of the Sun or Moon is a distant memory.

To identify a Nimbostratus, a cloudspotter needs simply to decide whether the cloud has a ragged, blurred dark grey base and whether the rain or snow that is falling from it is moderate to heavy and continual. If it does and it is, the cloud's a Nimbostratus.

But that's not to say that there isn't the possibility of confusion – there always is in cloudspotting. This might arise, for instance, were a cloudspotter directly below a Cumulonimbus. Like the Nimbostratus, this will have a dark base and, from directly below or if it is embedded within a layer cloud such as a Stratocumulus, it might seem to cover much of the sky. To confuse matters further, both clouds are often seen with accessory clouds, known as pannus. These dark patches of cloud can appear below the cloud bases when the air becomes saturated from all the precipitation.

The distinction can usually be made, however, by considering

Bob Norvill (member 58)

The Nimbostratus won't be winning any cloud beauty contests.

the nature of this precipitation. The showers from a thundercloud will be short-lived deluges. Even when one cloud is feeding into another to form a continuous 'multicell' storm, the showers will tend to be intermittent and will often contain hail. With more intense winds, a Cumulonimbus will also tend to be showing off with its telltale thunder and lightning.

A more likely confusion is with the Altostratus, from which the Nimbostratus so often develops. This is a shallower layer cloud, which can produce light precipitation. When it does so, as with the Nimbostratus, this is steady in nature.

When the Altostratus is of the thin, translucidus species, the distinction should be fairly straightforward. Its light grey underside will look far paler than the dark belly of the Nimbostratus.

But as a layer cloud thickens, it is a matter of judgement when to say it is no longer an Altostratus. Indeed, some say that it becomes a Nimbostratus as soon as precipitation falls. Thick enough, dark enough and wet enough, and a cloudspotter will be as correct as anyone else in calling it a Nimbostratus.

With all this rain and snow, the Nimbostratus is not exactly a dull cloud, but it does share with the Altostratus the distinction of being a cloud genus devoid of any species. Neither is it recognised as showing any varieties. Without any noteworthy variations in its appearance, the Nimbostratus is quite simply a thick, wet blanket, whose base is ragged and indistinct on account of its continually falling precipitation. It might be able to beat most of the other types in a fight, but it wouldn't get far in a cloud beauty contest.

☁

WHY DO SOME clouds produce rain, snow or hail, while others don't? If the fair-weather Cumulus is composed of water droplets, why is it a dry cloud, while the Nimbostratus is anything but?

This all has to do with the size of the water particles. Droplets and ice crystals all fall under the influence of gravity, but the smaller they are, the slower they fall through the air. With the exception of ground-based Stratus (fog or mist) all clouds form when air rises. Their particles only head groundwards if they grow large enough to descend through the air rising up below them.

A rainless, fair-weather Cumulus comprises very small water droplets – less than 0.005mm across. At this size, their speed of fall is insignificant compared with the up-welling thermal below. To their minuscule frames, the air is as dense as honey to a pebble.

Of course, a pebble dropped into a jar of honey will fall much more slowly than it will through a glass of water. But what if the pebble is dropped in as the jar is being filled with honey?

With the right size of pebble, and the right speed of filling the honey jar, it can be made to stay at a fixed position. That's what is happening when the water droplets or ice particles of a cloud seem *Falling* to disobey the laws of gravity – they are falling, but no *droplets and* faster than the air below is rising. (Important note: This is a *rising honey* hypothetical demonstration. Dropping pebbles into honey is a bad idea – especially at the breakfast table. They make the honey dirty and hurt the teeth when spread on toast.)

If this explains how cloud particles stay up, what makes them grow large enough to reach the ground?

☁

OR, 'WHY,' (as Frankie Lymon, the 1950s pop-star, sang) 'does the rain fall from up above?'

It's a good question – just one of the many that featured in Lymon's doo-wop hit, 'Why do Fools Fall in Love?' The song catapulted him into the US Top Ten at the tender age of thirteen. It reached No. 1 in the UK in the summer of '56 and made him the first black singer to sell a million records. Like many teen stars, Lymon had a troubled life, and I can't help thinking that it all had to do with the fact that he was a youngster with a lot of questions on his mind – questions that no one ever thought to answer. 'Why do birds sing so gay?' 'Why do lovers await the break of day?' 'Why do fools fall in love?' 'Why does the rain fall from up above?'

As the young Lymon toured America with his band, the Teenagers, his adoring fans queued up to buy the single. But did anyone ever think that he might have actually wanted to know the answers to his questions?

No, they didn't.

By the age of twenty, Lymon was a washed-up has-been with a drug habit, his only gigs nostalgia shows. The singer died, aged 26, on the floor of his grandmother's apartment from a heroin overdose.

How different things might have been.

If only someone had taken this troubled young man aside, sat him down with a cup of tea, and explained why the rain *does* fall from up above. He would at least have found the answer to one of his many questions. If The Cloud Appreciation Society and I had existed in the Fifties, I'd have gladly gone through it with him.

Would things have turned out differently for Frankie Lymon, had someone taken the time to explain to him why the rain *does* fall from up above?

Perhaps, if I'd caught him at the right moment, it might have changed the course of events.

Say I'd managed to talk my way backstage when the Teenagers were playing the London Palladium in 1957 and that I'd been able to hold his attention for ten minutes or so before he went out on stage. I'd have told him that rain or snow falls from a cloud when the water particles have grown a lot larger than the five thousandths of a millimetre across that you find in a young Cumulus cloud.

Cloud particles can grow much larger than that, and the greater their size, the more likely they are to hit the ground. Those that do can have a whole spectrum of sizes – from the finest 'Scotch mist' of a few hundredths of a millimetre across, through drizzle (0.2 and 0.5mm), to raindrops (greater than 0.5mm). Raindrops are typically between 1 and 5mm across. At sizes approaching the equivalent of 8mm, they become so distorted by the influence of air resistance that they break into smaller droplets.

By now, Lymon's interest would have been piqued. But I don't think this alone would have answered his question fully. He'd need more than the basics.

He'd need to know *how* the droplets in a cloud grow large enough to fall from it.

If my 'explanation intervention' was to have any success, I'd have to hope that he had a few minutes more before his set began. I'd tell the groupies, the tour managers and all the other hangers-on to take their Jack Daniels out of the backstage area and leave us alone. I'd need to introduce Lymon to the two processes by which rain forms in clouds – both the one that happens in the parts of clouds made of water droplets, and the other in those of ice. To introduce the first, I'd talk to him about pearls and oysters.

☁

HINDU MYTHS HELD that a pearl forms when a drop of dew falls into the sea. If it happens on a full moon, they said, the pearl is a perfect one. Ancient Greeks had a different explanation. They claimed that a pearl is created when lightning strikes the sea. And as for the Romans, it was just a mermaid's tear. The present-day explanation is rather more prosaic: pearls form when a little piece of grit gets inside an oyster's shell. The crustacean has a *Pearls and* gland that secretes mother-of-pearl to coat the shell's *raindrops* interior. When grit is present, this acts as a nucleus onto which the mother-of-pearl starts to accumulate. It takes about a year for the finished pearl to form. By this time, the oyster will be feeling decidedly cramped and will tend to cast its unwanted treasure into the deep.

Grit of an airborne kind is also needed for water vapour to condense into a cloud. The free-flying, individual molecules of water will only tend to combine into liquid droplets if there is something for them to get started on. They need a nucleus onto which they can collect. In fact, there are plenty of 'pieces of grit' floating around in the atmosphere that can serve the purpose. Meteorologists call them 'cloud condensation nuclei'. They are generally less than 0.001mm across and can take many forms.

Over the sea, they can be particles of salt from dried-out sea spray. Over land, where they are much more plentiful, they can take the form of dust particles of clay, minerals or dried-up vegetable matter. Soot from volcanoes and forest fires will also do the trick; so will the huge amount of combustion products such as smoke and acid particles near human habitation. These condensation nuclei are crucial to the first way that rain is formed – in clouds that contain liquid water droplets alone (being too warm for their droplets to freeze).

It turns out that condensation nuclei vary in how efficient they are at attracting the molecules of water vapour. Some happen to hold on to them better than others. As anyone who has tried using a saltcellar in humid weather will confirm, salt particles are particularly good at absorbing water molecules. These act, therefore, as very efficient cloud condensation nuclei.

Some of the particles from fires aren't bad either, which is one reason thick 'pyrocumulus' clouds can appear over forest fires.

As a cloud begins to appear, those droplets that take form on the more efficient nuclei grow more rapidly than the others. Over time, they become larger and so start falling faster. When they are large enough, they begin colliding with the smaller droplets, and growing as they do so. Meteorologists call the process 'coalescence'. It is one way in which, after some 15–30 minutes of its existence, a cloud can develop droplets large enough to fall as rain. But (I would tell Lymon) it's by no means the most common way – at least, not in the mid-latitude regions of the world.

☁

BY NOW, THE PEOPLE backstage would be calling to Frankie that he was due on stage any minute. But I'd need a few moments more with him to explain the main way that raindrops develop. To explain that, I would need to bring in ice crystals. Much more often than not, the rain that reaches us down on the ground starts life in solid form. It only melts into raindrops as it passes through the warmer air below the cloud. To explain the second process by which rain falls, I would need to tell Lymon a thing or two about the peculiar way that cloud droplets freeze into ice crystals.

On a typical autumn day over Britain, the air temperature might drop to 0°C at around 6,500ft above the ground. You might think that the droplets in any cloud above this altitude would therefore freeze. But one of the surprising things about clouds is that their droplets don't freeze at 0°C. In fact, they don't typically freeze until temperatures become a lot colder than that.

Whilst a puddle of water down at sea level might start freezing at 0°C, droplets suspended in the atmosphere behave rather differently. You see, just as 'pieces of grit' are needed for water *Incredible* vapour to condense into airborne droplets, nuclei are also *unfreezable* needed for those droplets to freeze into solid crystals (at *droplets* anything above very cold temperatures). And it turns out that water – whether in liquid or gaseous form – is a lot fussier about the type of nuclei it freezes on to than the ones that it condenses on to.

Atmospheric particles that serve as 'icing nuclei' are much larger than condensation ones. At 0.005 to 0.05mm across, they have 100 to 130,000 times larger volumes. Primarily consisting of rock or other mineral fragments, icing nuclei are also a lot less common.

There just aren't that many of them around in the atmosphere, and without something to get started on, a droplet will stubbornly refuse to freeze until it reaches temperatures as low as −35 to −40°C. Without the right size and shape of nuclei, the droplets will remain in what is called a 'supercooled state'.

Lymon's backing band would be out on stage by now, his manager banging on the door. I wouldn't have long to wrap things up. But I would be coming to the crux of my explanation.

A cloud with depth, like the Nimbostratus, can have a large proportion of supercooled droplets in its cold upper reaches. The higher these droplets rise, the colder they become. Without the right nuclei, however, they are in no hurry to change state − cooling to −5°, −7°, −10°C, they stubbornly refuse to freeze. Like prima donna musicians refusing to perform because the M&Ms backstage are the wrong colour, droplets can remain as supercooled liquid until temperatures are in the −20s.

When it becomes that cold, they become decidedly less fussy. Now particles that are roughly the right size and shape will be enough to encourage freezing. It is as if a few of the prima donnas say, what the hell − we're OK with these M&Ms, and agree to go on stage. No sooner have some of the droplets started to freeze on to the more plentiful 'just OK' nuclei than − would you know it − suddenly they're all at it.

The first few droplets start to freeze from the outside in − forming, at first, a solid shell with a soft centre. As anyone whose pipes have burst in the winter will appreciate, when the centres freeze too, they expand and crack the outer shell. Little fragments of ice, or 'ice spikules', burst out and break off. These then act as icing nuclei for other droplets to freeze on to and a chain reaction takes place of the tiny supercooled droplets freezing.

Being solid, ice crystals hold on to water molecules more tightly than the droplets do. This means that they soon grow in size as molecules migrate towards them from the remaining supercooled droplets. Soon, the crystals are large enough to pick up considerable fall speeds. And as they start descending, they collide with more supercooled droplets which freeze on to them, growing them still further. Soon they come falling out of the bottom of the cloud and, in the warmer air below the cloud, they can melt again and land as rain.

That, more often than not, is how the rain falls from up above.

I'd have finished just in the nick of time – moments later Frankie Lymon would run out on stage to the roar of his adoring fans. In the rock and roll mêlée, as he was ushered down the corridor, I'd just have time to call out after him: 'Frankie… It's known as the Bergeron–Findeisen process!' (for they were the guys who worked out how ice crystals grow large enough for the rain to fall from up above).

Would my explanation intervention have saved young Lymon? Perhaps it might just have been enough for him to put those confounded questions behind him and live a full life after all.

WHEN CHILDREN DRAW RAINDROPS, they invariably give them the shape of tears. I suppose that is how adults teach them to do it – like the way they learn to draw a Christmas tree with its branches pointing diagonally downwards.

But just as a Christmas tree actually has its branches pointing upwards, raindrops don't fall in the shape of tears. Tiny cloud droplets may be pretty much perfect little spheres but, once they've grown large enough to fall fast, they are greatly distorted by air resistance and are not shaped like spheres – or indeed teardrops – at all. When they are a couple of millimetres or more across, raindrops actually look like the top half of hamburger buns.

It would be a bit much to expect children to draw tiny hamburgers falling from their clouds (though I'm sure there is a McDonald's marketing campaign in it). No doubt, we teach them the convention of tear-shaped raindrops because we are so used to seeing drops in that shape when they fall from objects – like the bath tap. Anyone who watches drips of water clinging to the tap rim will know that, as they start to fall, they *do* look like tears. They extend under their growing weight, clinging with all the might of their surface tension, only to finally lose themselves in the bathwater below.

Just as it is wrong to draw Christmas trees with the branches pointing downwards, it is also wrong to draw raindrops in the shape of tears. Children who insist on doing so should be severely punished.

Of course, we see them falling in that way from our loved one's eyelash. And so perhaps we are also lead to depict raindrops as tears because of an association between rain and sorrow. It is an association that was beloved of the English Romantic poets. John Keats drew upon it in his somewhat depressing poem 'Ode on Melancholy':

> ... when the melancholy fit shall fall
> Sudden from heaven like a weeping cloud,
> That fosters the droop-headed flowers all,
> And hides the green hill in an April shroud;
> Then glut thy sorrow on a morning rose.[2]

Percy Bysshe Shelley did so too when he wrote 'Adonais' a year later, on the news of Keats's death. The eulogy to his departed friend was also a barrel of laughs. Shelley imagined how the 'quick Dreams' of Keats's imagination would die with him. At his deathbed, one of these (a somewhat amorous lady Dream) would see a tear on his cheek. Hoping that he hadn't died after all, she'd cry out:

> *'Like dew upon a sleeping flower, there lies*
> *A tear some Dream has loosened from his brain.'*

But, unfortunately for her:

> *She knew not 'twas her own; as with no stain*
> *She faded, like a cloud which had outwept its rain.*[3]

CLOUDSPOTTERS WHO THINK that precipitation is either rain, snow and hail had better think again. These are just three of the many forms that it can take. It is time for a more exhaustive list of what can fall from clouds. The different forms depend on the type of cloud and the air temperatures within and below it:

LIQUID PRECIPITATION:
RAIN: drops of water that are generally larger than 0.5mm in diameter.
FREEZING RAIN: supercooled drops (at temperatures below 0°C), sometimes called sleet, which are liquid water at temperatures below 0°C, and tend to freeze on contact with the ground or objects, such as telephone cables.
DRIZZLE: falling drops of water, which are closely spaced and very fine, being generally less than 0.5mm in diameter.
FREEZING DRIZZLE: supercooled drizzle (at temperatures below 0°C), which, being in the form of smaller droplets than freezing rain, can remain liquid at even lower temperatures. This can cause a lot of damage to property and aircraft by freezing as 'rime' instantly on contact.

	Rain	Drizzle	Snow	Snow grains	Snow pellets	Hail	Small hail	Ice pellets	Cats & dogs
Cumulus	●		✳		△				
Cumulonimbus*	●		✳		△	▲	△		🐱
Stratus		❟	✳	⊖					
Stratocumulus	●		✳		△				
Altostratus	●		✳					⊿	
Nimbostratus*	●		✳					⊿	
* Genera that are, by definition, precipitation bearing.									

Most clouds don't cause precipitation. But here is a chart of those that can, and anyone who thinks they only produce rain or snow had better think again.

SOLID PRECIPITATION:

SNOW: crystalline formations of ice. It can either take the form of single crystals or clumps of entangled crystals, known as snowflakes (usually at temperatures above –5°C). The shape, size and concentration of the crystals can vary greatly, depending on the temperature and conditions in which they form.

SNOW GRAINS: known as 'graupel', these are very small opaque white crystals of ice (generally less than 1mm in diameter), which do not bounce on hitting the ground. They are a little like the snow equivalent of drizzle.

SNOW PELLETS: white, opaque ice crystals, which are generally conical or rounded and are between 1 and 5mm in diameter. They are usually brittle and are easily crushed. Falling on hard ground, they bounce and often break up. Snow pellets form as a result of ice crystals, such as snow grains, colliding with cloud droplets, which freeze as solid blobs on the outside.

HAIL: very hard ice particles that tend to have diameters of between 5 and 50mm (though the largest recorded in the USA was 178mm wide). They can be either clear or opaque and are generally observed during heavy thunderstorms, building within the cloud in layers of ice, like a frozen gobstopper.

David Foster (member 1157)

Even if it is rather a wet blanket, Nimbostratus – like the other precipitating clouds – performs an essential role in cleansing the atmosphere of pollutants.

SMALL HAIL: translucent ice particles, generally with a diameter of less than 5mm, which are hard to crush, fall in showers and bounce with an audible sound on impact.

ICE PELLETS: transparent ice particles, with a diameter of less than 5mm. Also hard to crush and making a sound when they hit the ground, these tend to fall in a more prolonged and steady manner than small hail.

DIAMOND DUST: very small ice crystals (usually around 0.1mm in diameter), which often appear to be suspended in the air. This dust forms in clear, calm and very cold air, as might be found at the poles. It is the one form of precipitation that does not fall from a cloud and gets its name from the beautiful way that it sparkles in the sunlight.

I can't imagine that Frankie Lymon knew these less familiar forms of precipitation. Perhaps that is just as well, as he wouldn't have gone far singing 'Why does the graupel fall from up above?'

☁

CLOUDSPOTTERS OFTEN FIND themselves having to defend our fluffy friends in discussions with unenlightened acquaintances. 'How can you possibly like miserable rain clouds?' the latter will say, moving on to moan about how rain clouds a) delayed yesterday's tennis match, b) ruined their wedding, c) caused devastating floods in Bangladesh... and so on.

For all the dampening of spirits – and worse – blamed on the rain, it is worth pointing out that without the clouds' role in desalinating the oceans, there wouldn't be anything to drink. In the words of St Basil the Great, from the fourth century:

> *Many a man curses the rain that falls upon his head, and knows not that it brings abundance to drive away the hunger.*

Or as John Updike, the American novelist, put it: 'Rain is grace; rain is the sky condescending to the earth; without rain, there would be no life.'[4]

But there is another equally critical role that clouds and precipitation play in making Earth habitable – one that is perhaps less obvious. They are one of the primary ways that the air is cleaned of pollutants.

Cloud condensation nuclei and icing nuclei are trapped within the cloud's particles, and are returned to the surface with its precipitation. Only 2.5cm of rainfall is enough to remove around 99 per cent of airborne particles and almost all soluble gases, such as sulphur dioxide, from the part of the atmosphere below. Precipitating clouds carry the atmospheric nuclei, on to which their droplets and crystals formed, back down to Earth. Without them, the atmosphere would be unspeakably hazy, acrid and – certainly in temperate regions of the world – deadly.

Those who complain about the rain are blind to the big picture. There are few things more invigorating than the clear, fresh air after rainfall. As any cloudspotter knows, sunshine is so delicious because it breaks through, as the cloud weeps away its rain.

Might those hamburger-shaped tears, in fact, be ones of joy rather than sorrow?

The High Clouds

EIGHT

CIRRUS

The delicate streaks of
falling ice crystals

O f all the common clouds, Cirrus must be the most beautiful. Their name comes from the Latin for a lock of hair, for they are the delicate bright white wisps of ice that appear high in the heavens. Joni Mitchell, the hippy Canadian songwriter, likened them to floes of angel hair in her 1969 song, 'Both Sides Now'. Presumably, this is after the angel has used the most heavenly conditioner.

It is tempting to compare Cirrus to the pale streaks across slabs of marble, or the fine traces of fat in a quality side of beef, but those comparisons are far too solid. A more fitting description involves Frigga, the Norse goddess of the atmosphere, who would wear either bright white or dark garments, according to her somewhat variable moods. Her palace, Fensalir, contained a hall of mists – something that every home should have – and here Frigga would sit with her jewelled spinning-wheel to weave long webs of cloud. This is how Cirrus clouds look – as if spun from the finest celestial silk, with a 'Made in Fensalir' label attached.

☁

CIRRUS ARE THE HIGHEST of the common clouds and are composed entirely of ice crystals, typically forming above 24,000ft in temperate regions of the world. 'Pencilled, as it were, on to the

HOW TO SPOT
CIRRUS CLOUDS

Cirrus are the highest of the ten main cloud types. In the form of delicate, white streaks, patches or bands of falling ice crystals, they are detached from each other, and have fibrous or silky appearances. Cirrus rarely appear very thick. They are often seen with the other high clouds, Cirrostratus and Cirrocumulus and, like them, can show 'halo phenomena' around the Sun or Moon.

TYPICAL ALTITUDES*:
16,500–45,000ft
WHERE THEY FORM:
Worldwide.
PRECIPITATION (REACHING GROUND): None.

Cirrus floccus

Cirrus uncinus

Cirrus vertebratus

CIRRUS SPECIES:

FIBRATUS: When it is in the form of straight or curved filaments that are mostly distinct from each other and do not terminate in hooks or clumps.

UNCINUS: When its 'fallstreaks' are the shape of hooks or commas.

SPISSATUS: The thickest Cirrus – when it is in patches that appear grey in front of the Sun – which tends to originate from the anvil of a Cumulonimbus.

CASTELLANUS: When it is in the form of small distinct clumps with crenellated tops.

FLOCCUS: When it is in the form of independent small round tufts, which often show trails of ice crystals falling from them.

NOT TO BE CONFUSED WITH...

CIRROSTRATUS: which looks like a thin, milky smooth or fibrous veil across the sky. Cirrus, by contrast, is in separated streaks, fibres or patches.

CIRROCUMULUS: which is a high layer of cloudlets, like grains of salt. Cirrus does not show this finely dappled texture.

CIRRUS VARIETIES:

INTORTUS: When the fallstreaks are irregular and tangled.

RADIATUS: When the filaments are in parallel bands, usually aligned to the wind at high altitude, which converge towards the horizon, due to perspective.

VERTEBRATUS: When the filaments look like a fish skeleton.

DUPLICATUS: When the filaments, streaks or hooks are arranged at more than one altitude, which can be apparent when the winds cause them to point in different directions.

* These approximate altitudes (above the surface) are for mid-latitude regions.

So that's how Cirrus are formed.

sky' is how Luke Howard described them, and their delicate tendrils seem all too often ignored down on the ground. Perhaps this is because they are not responsible for any noticeable change in the weather – they don't produce rain or snow (at least none that reaches the ground), nor are they associated with ground-level winds. Moreover, they barely seem to weaken the daylight, since they are rarely thick enough to block out much of the Sun.

You wouldn't think it, but Cirrus are in fact precipitating clouds. The reason they are not defined as such is that their 'precipitation' evaporates away in the warmer air below the cloud. Nevertheless, to look up at Cirrus clouds is to see snow – well, ice crystals, to be precise – falling too high to reach the ground. Cloud-spotters may live in regions too warm for snow but, in the Cirrus clouds, they can still see how it looks from a few miles off.

The movement of the falling crystals, as they are whipped by the winds high in the troposphere, is what gives Cirrus their distinctive, flossy formations, known as 'fallstreaks'. Up at the top of the troposphere, winds will often be as strong as 100–150mph. Forming in gales like this, Cirrus aren't clouds that stick in one place for long.

Compared to low Cumulus gliding along in the breeze, these clouds can look almost stationary. Cloudspotters should remember, however, that the further away something is, the more difficult it is to notice its motion. In fact, Cirrus move a lot faster than Cumulus clouds. Not only are they the highest of the common clouds, they are also the fastest.

On average, wind speeds decrease with descent through the troposphere. So, as the ice crystals of a Cirrus cloud fall, they tend to pass into slower air currents below, where they lag behind. The varying temperatures and humidities on the way down can cause the crystals to grow and multiply in some regions and diminish in others. It is this variation in the wind speed, temperature and

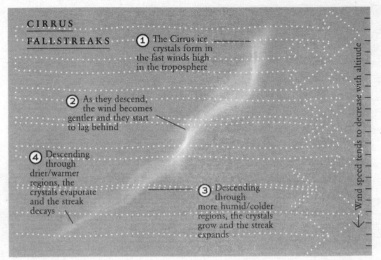

The wind speed, air temperature and humidity determine the appearance
of the Cirrus cloud's fallstreaks.

moisture content of the air through the crystals' descent that
determines the wavy forms of the Cirrus fallstreaks.

How could these silent wafts of ice hope to compete for our
attention with clouds like the roaring Cumulonimbus? They are
just too serene to distract us. But a cloudspotter will be wise to pay
careful attention to these clouds. Although they may not lead to
noticeable weather down on the ground, in their understated way
they can tell us a great deal about the weather in store. They can be
Nature's way of saying that there are changes afoot. To those who
know how to listen, they are little whispers that warn of the
approach of more weighty clouds – ones that certainly *do* make a
difference to weather down here.

That Cirrus can be harbingers of a 'deterioration' in the weather
only adds to their fragile beauty – are not the most delicious things
the ones we know can't last?

☁

THERE ARE FIVE different species of Cirrus, distinguished by the
appearance and orientation of the cloud's streaks. Sometimes they

are in the form of elongated filaments, either straight or slightly curved, which are generally distinct from each other, and do not end in tufts or blobs. In this case, the cloud is known as Cirrus fibratus. The species commonly referred to as 'mare's tails', and officially called Cirrus uncinus, is when the streaks are shaped like commas or hooks. The top of each comma is thicker than the tail but, even on careful inspection, does not show any clumps or mounds on its upper side.

Cirrus spissatus is when the cloud is thick enough to appear as grey patches when viewed towards the Sun. This species is often formed when the anvil of a Cumulonimbus remains after the rest of the storm cloud has dissipated. By contrast to spissatus, the other species are usually transparent enough to always appear bright white.

The species known as castellanus is, as with the Altocumulus and Stratocumulus clouds, when the individual Cirrus clouds have turret-like protuberances rising from a common base. Finally, floccus is when the Cirrus is in the form of small, rounded tufts, often with trails falling below them.

With all the species, it is important to remember that Cirrus are more or less independent clouds. If the cloud elements are joined into a flat veil or into a layer of tiny regular cloudlets, then it is one

Mike Davies (member 1633)

Cloud genera don't have to be defined as one of the species. These are just Cirrus.

of the high layer clouds called Cirrocumulus or Cirrostratus. All three types seem fond of each other's company, as it is common to see them at the same time.

Cirrus can exhibit four different varieties. When its filaments are twisted and tangled, the cloud is of the beautiful, and sometimes a little unsettling, variety called intortus. Vertebratus, on

Cirrus and fish – on or off the bone

the other hand, is when they are lined up regularly like the skeleton of a fish. (This is not to be confused with the mackerel sky of the Cirrocumulus cloud, which is composed of bands of tiny cloudlets, which look like fish scales. Cirrus vertebratus is after the fish has been eaten.)

Radiatus and duplicatus are varieties shared with other cloud types. Radiatus refers to when lines of the Cirrus appear to converge towards the horizon (as a result of perspective), while duplicatus is when the Cirrus are at two or more quite distinct altitudes. The latter can be rather difficult to identify, since it is hard to judge the altitudes of such ethereal wisps and also because the cloud is composed of crystals that are falling down considerable distances anyway. Spotting a duplicatus becomes easier when the wind direction differs with the Cirrus levels, so that the filaments at one level are in a different orientation to those at the other. Sunrises and sunsets also help, since the lower layer can be in shadow – and so appear grey – while the other remains lit by the low Sun.

☁

WITH CHAPTERS ON EACH of the different cloud types in turn, I rather run the risk of portraying them as quite separate and distinct beasts. In reality, nothing could be further from the truth. Clouds are in a constant state of flux – forever transmuting from one formation into another, as they reflect the changes in atmospheric temperature and humidity.

Nowhere is this more apparent than in the temperate middle latitudes. It is these regions, half-way between the Equator and the poles in both the Northern and Southern Hemispheres, which experience the most changeable weather on the planet. Consequently, they present the greater challenges to forecasters.

Jeffroy Rathbun (member 1590)

This rare Cirrus formation is known as a Kelvin–Helmholtz wave cloud and can form in the region between shearing winds, moving in different directions.

In spite of the capricious behaviour of clouds in these latitudes, they do sometimes change in fairly predictable sequences; and there is one pattern that begins with the appearance of thickening Cirrus clouds, with which every cloudspotter should become familiar. It often takes a day or two for the sequence to be complete but, if they notice it at an early stage, cloudspotters can anticipate the unpredictable weather of the temperate regions and learn to recognise the atmosphere's changing moods. Few things can be more important to a cloudspotter. It might feel at first like learning Latin nouns, but it will be worth it in the end.

THE SEQUENCE STARTS with the appearance of the Cirrus's faint ice-crystal wisps over generally clear skies, except perhaps for a few low-level Cumulus. The Cirrus gradually spread out across the sky, merging together. How they spread is worth noting, for it gives an indication of the direction of the wind at the top of the troposphere. When this is blowing at a right angle to the wind at

Graham Tilston (member 562)

Cirrus floccus consists of independent tufts of high cloud,
with trails of ice crystals falling below them.

ground level, it can be a clue that the classic sequence is beginning. In the Northern Hemisphere, if you stand with your back to the wind and the Cirrus are spreading out to your right, you can take this as evidence that the sequence is beginning – evidence that the clouds are about to change in a way that indicates the passing of a region of low pressure, or a 'depression'.

Before long, the spreading wisps of Cirrus will start to join together, shrouding the sky in a milky veil, which will flatten out and lose its definition. Soon, they will no longer be Cirrus, but will have changed into the high layer cloud of crystals called Cirrostratus. This will never be thick enough to block out much of the Sun but, as it thickens, it lowers its base towards the ground, developing into the mid-level Altostratus, which usually contains droplets as well as crystals.

The transition to Altostratus will make the Sun appear as if seen through ground glass, the sky now looking overcast, with a flat light grey colour. This Altostratus layer will gradually thicken further, its base lowering still, and its shade changing from a bright to a duller grey. By now, even those blind to the shifting moods of the sky will be saying rain is likely and, sure enough, small drops of rain – or snow, if temperatures are low enough – will begin to fall in a steady, light way.

Soon, the base of the cloud layer may have descended to only 1,000ft. Now dark and heavy looking, the cloud has changed into the Nimbostratus rain cloud, its precipitation falling more heavily. The rain or snow will be prolonged and steady, and just as it was slow to begin, so it will be slow to finish. After some hours the Nimbostratus will have begun to brighten as it thins out. Once it has rained itself out, it will have transformed into a low Stratus layer, which finally breaks up into a Stratocumulus and perhaps clears once again to leave individual Cumulus clouds.

This, the first part of the sequence, typically takes place over a day. It tends to be accompanied by an increase in air temperatures at ground level. It is formation dancing performed by the layer clouds and, as is the way of these cloud types, it doesn't happen in a hurry – just a gradual and prolonged progression, with gradual and prolonged precipitation.

In summary, the classic sequence accompanying a rise in air temperatures goes as follows:

SPREADING CIRRUS (perhaps with some Cumulus around and about)
CIRROSTRATUS
ALTOSTRATUS (gradually leading to light rain)
NIMBOSTRATUS (leading to prolonged and heavier rain)
STRATUS
STRATOCUMULUS
CUMULUS (and clearer skies once again)

The progression typically takes up to a day

In the second half of the sequence it is not the layer clouds that are the stars of the show, but the individual convection clouds. This means that the dance is rather more frenetic.

With a drop in the air temperature again, a different and more sudden pattern of clouds can occur. Often this part will begin with the appearance of an Altocumulus layer, or perhaps the higher, ice-particle equivalent known as the Cirrocumulus, known by mariners as a 'mackerel sky' for the way it can resemble the scales of the fish.

The sailor's saying, 'Mackerel sky, mackerel sky, not long wet, not long dry', is an indication of what is to come. As the temperature drops, Cumulus congestus or even Cumulonimbus storm clouds can dramatically grow, building from energetic Cumulus clouds or out of layers of Stratocumulus. Of course, these mountains of moisture never hold their water for long, and the sudden showers of rain, snow or hail that fall from them, although often heavy, are usually brief. The convection cloud flare-ups tend to rain themselves out just as fast as they formed, and the winds they whip up can be fierce.

With the passing of the storm, all that may be left are patches of Altostratus and, with a pleasing symmetry, the very same high wispy Cirrus with which the whole pattern began. These residual Cirrus can hang around for some time and any cloudspotter who has hitherto been preoccupied with tedious matters on the ground, and just thought to look up, might wonder if the layer-cloud sequence is only now beginning. Are they watching the start of the performance, or the closing moments?

The wind directions will help to tell them that the show is winding up. With their back to the wind at ground level, cloudspotters in the Northern Hemisphere will now notice that the Cirrus are being blown to their left, not their right, as they were at the start.

The second half of the cloud sequence, accompanying a drop in air temperatures, therefore, goes something like this:

ALTOCUMULUS or CIRROCUMULUS (the sailor's mackerel sky)
CUMULUS CONGESTUS or CUMULONIMBUS (bringing sudden and heavy showers)
ALTOSTRATUS (in patches, left behind by the storm clouds)
CIRRUS

The process typically takes a matter of hours

The sequence is in two halves because they sometimes occur independently. They also vary considerably in intensity – sometimes being too gentle even to develop the precipitating clouds.

If that did feel like a lesson in Latin nouns, cloudspotters will find the sequence of clouds associated with the passage of a depression a lot easier to remember if they understand why they develop in this way.

☁

THIS WAS WORKED OUT just after the First World War by a brilliant group of meteorologists in Bergen, on the southwest coast of Norway. Bergen is one of the wettest towns in the whole of Europe, with an annual rainfall of 88 inches that makes London's 22 look rather paltry – no doubt the Bergen School, as these meteorologists became known, had more reason than most to ponder the development of rain clouds.

Not only did they come up with a way of understanding why clouds in temperate regions tend to have this sort of progression, they also found out something a lot more fundamental. The Bergen School came up with a way of making sense of the weather in the world's temperate zones that was one of the most outstanding contributions to our understanding of the changeable weather of the middle latitudes.

Their work provided a major breakthrough in forecasting in such regions. More importantly, it helps cloudspotters remember the typical cloud sequence, which begins with the spreading of the Cirrus cloud's high, wispy, ice-crystal streaks.

The Bergen School was established by a weather boffin called Vilhelm Bjerknes when he moved to the city in 1917. Bjerknes was returning to Norway after five years at the University of Leipzig. Throughout the First World War, the crucial business of forecasting the weather had been primarily a matter of studying changes in the air pressure shown on barometers. Ever since the invention of the barometer in Florence in 1644, it had been known that when the air pressure drops, clouds, and so rain, become more likely. Despite the urgency of improving weather predictions, no one really knew why changes in the weather are associated with changes of the air pressure, as shown on a barometer.

On returning to Bergen, Bjerknes found his country on the brink of famine. Its cold climate and rocky terrain meant that Norway had traditionally relied heavily on imported cereal grains for its food. The war had disrupted supplies to such an extent that emergency measures were implemented in an attempt to increase agricultural production, and Bjerknes was given the job of reorganising the country's weather service to provide essential information for the struggling farmers.

They needed as much warning as possible of the arrival of storms that could damage their fragile crops. This was motivation enough for Bjerknes and the group of young meteorologists that he gathered around him to find out why regions of low pressure are associated with rain and storms. In 1918, they realised that it is not the change in air pressure that causes the systems of wet and stormy weather in the middle latitudes. Both the pressure changes and the rainy weather result from extensive areas of warm air and cold air coming into contact with each other.

The Bergen School was the first to propose that the atmosphere behaves like an enormous heat engine. The parts above the hot tropics are warmed whilst those above the poles remain cold. In an attempt to even out the temperature difference, air moves around the globe redistributing the heat. The Norwegians discovered that this movement can be considered in terms of distinct air masses, which, like warm and cold ocean currents, actually mix less than you might imagine.

They discovered that there are, in fact, wavy lines of temperature 'discontinuity' – where regions of warm air, originating from the Equator, come up against colder ones from the poles. They found that this discontinuity actually winds around the globe at latitudes of around 50–60° in both the Northern and Southern Hemispheres.

It is along this meandering, shifting boundary of air that the changeable weather of the temperate zones is born. With the war fresh in everyone's mind, Bjerknes dubbed the line of conflict the 'polar front', borrowing the term from combat parlance. It was an

The spreading and joining of Cirrus clouds in the middle latitudes
can be early signs of an approaching 'depression'.

appropriate name, for these fronts are where skirmishes break out
between differing air masses. They are also where weather systems
develop that lead to the common cloud progressions that begin
with the Cirrus whisper and end with crashing Cumulus congestus,
or even the roar of the Cumulonimbus.

With the ebb and flow of competing air – warmer air moving
from the tropics towards the poles, and colder air doing the
opposite – the border of this polar front is constantly snaking
northwards and southwards, blurring and sharpening, breaking and
reforming. It is an ever-changing line of battles, all of them fuelled
by no more than the difference in the Sun's warming effect on the
various regions of the globe.

It wasn't until the Second World War that pilots noticed
extremely fast ribbons of wind high in the troposphere, which it
turned out correspond to the general position of the polar fronts.
These high-speed 'jet streams', as they became known, course from
west to east high in the troposphere above the meandering lines of
discontinuity. Up at around 30,000ft, pilots flying against a jet
stream were surprised to find they were virtually at a standstill. Yet

Keith Epps (member 868)

Dr Malcolm Buck (member 1170)

Who cares if Cirrus clouds are heralding a 'deterioration' in the weather
when they look as beautiful as this fibratus species?

with the stream behind them, their journey times could be slashed
dramatically. The winding ribbons of the jet stream above the
Northern Hemisphere polar front is why a flight from London to
New York can take an hour longer than the return journey.

Sometimes 'jet-stream Cirrus' can be seen stretching in long
streaks off to the horizon, as their ice crystals are swept along in
these rivers of high-altitude, high-speed winds. Jet streams also have
a considerable effect on the movement of any kinks and ripples
that develop along the polar front.

It is with the movement of these kinks in the border between
air masses, as they are propelled eastward by the jet stream, that the
Bergen School began to make sense of the repeating patterns of
cloud formation found in the temperate zones.

☁

WITH THE GROWING ACCEPTANCE of the Norwegians'
model, meteorologists began to shift their gaze from their
barometers up to the clouds. For the first time, there was a way of
making sense of the relationship between developing cloud
formations and approaching weather in temperate regions. Whilst

the idea that particular cloud formations presaged storms had been observed since at least as far back as Aristotle's time, scientists had generally considered it to be little more than folklore. Bjerknes and his assistants found that particular formations appeared with surprising consistency at the collision of warm and cold air masses.

A cloudspotter in the middle latitudes can notice distinct changes in the air temperature as a pronounced kink in the polar front is swept overhead by the effect of the jet stream. First they'll feel the cooler (polar) air being replaced by warmer (tropical) air, which is then replaced by cooler air once again as the kink passes. It is along the boundaries of differing air masses that clouds, and so precipitation, can develop.

WARM AIR IS LESS DENSE than cold, so it tends to ride up over the top when they come into conflict. If the warmer air has crossed oceans on its journey from the tropics, it will have picked up a lot of water vapour *en route*. By cooling as it rises over the cooler air, some of this can condense into clouds.

The rising, warmer air causes barometers to drop, indicating that air pressure at ground level is decreasing. The drop in pressure as the warmer region rises is why the 'battle of air' is known as a depression.

The two halves of the classic cloud sequence occur at the temperature boundaries of the passing depression. The more marked the differences between the temperatures and humidities of the air masses, the heavier the cloud and precipitation.

The more gradual, layer-cloud sequence appears with the arrival of the warmer air (initally, at the upper levels of the troposphere). The more sudden, convection cloud sequence follows as the warmer air is pushed ahead again by the colder air. The kink in the line of discontinuous air masses therefore consists of two local fronts, known as 'warm' and 'cold' fronts.

The gradual incline of the warm front at the fore of the kink causes a gentle lifting of large regions of the warm air, leading to thickening high Cirrus wisps followed by deeper and deeper layers

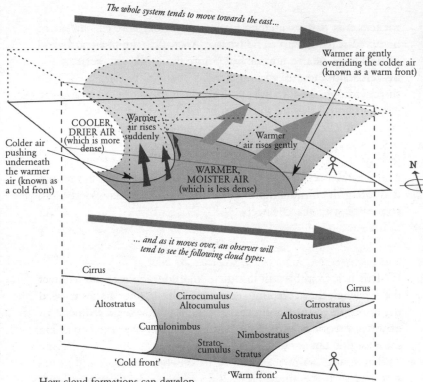

The whole system tends to move towards the east...

Warmer air gently overriding the colder air (known as a warm front)

COOLER, DRIER AIR (which is more dense)

Warmer air rises suddenly

Warmer air rises gently

Colder air pushing underneath the warmer air (known as a cold front)

WARMER, MOISTER AIR (which is less dense)

N

... and as it moves over, an observer will tend to see the following cloud types:

Cirrus

Cirrus

Altostratus

Cirrocumulus/ Altocumulus

Cirrostratus

Altostratus

Cumulonimbus

Nimbostratus

Strato-cumulus   Stratus

'Cold front'

'Warm front'

How cloud formations can develop
as a region of low pressure, or 'depression', passes over.
Those who think this looks complicated will be depressed to learn
that it is in fact a very simplified diagram of a weather system.

of cloud. The skies can clear between the two halves of the cloud progression, since the region of warmer air is not being lifted. But with the arrival of the cooler air again, this soon changes to the second part of the sequence. Here, the warmer air rises suddenly as the cooler burrows beneath. The volatile Cumulonimbus clouds can form along the cold front – building dramatically in enormous towers. All the sudden lifting also results in strong winds.

The Bergen School's model of interacting air masses provided a vindication for all those who saw the clouds as faithful indicators of the weather ahead. Luke Howard likened the clouds to expressions on a person's face. 'They are commonly as good visible indications of the operation of these causes [of the weather],' he wrote, 'as is the countenance of the state of a person's mind or body.'[1] The gently spreading wisps of Cirrus can be the first flickers of a change in the atmosphere's mood. It may have taken the Bergen School to explain it, but this is something that observers of the clouds had long known: 'See in the sky the painter's brush, the wind around you soon will rush.'

☁

THE LEAP IN OUR UNDERSTANDING of the weather that was derived from considering the movement of competing air masses, combined with the perspective of satellite imagery and the power of high-speed computers, has helped to improve forecasting hugely over the last fifty years. We have learnt to rely on the media more and more to inform us of the weather ahead.

This is a great help in deciding whether to plan a barbecue at the weekend, but it also means that we are forgetting how to read the atmosphere's changing moods. Whilst we are able to watch its expressions, as manifested in the clouds, we are becoming increasingly ignorant as to their meanings. It is as if we are becoming meteorologically autistic.

In 1156, the Chinese author Yeh Meng-te wrote: 'Since I had plenty of leisure time, I usually rose early in the morning, and then with an empty mind concentrated on the beauty of the fields, trees, rivers, mountains and clouds and I found that I could predict the weather right seven or eight times out of ten. Then I realised that in quietness the universe can be observed, the inner moods felt and real truth obtained.'[2]

I loved it when I first followed the classic progression of clouds that begins with spreading Cirrus. I was on a train travelling southwest from London. The journey meant that I was moving into the advancing weather system, so I could watch the clouds develop more quickly.

Those who hate clouds and precipitation might consider that a 'depression' is a fitting description for the advance of a region of lifting, warmer air, but I find it anything but. It was an April day – the time of year in Britain when the clouds seem at their most active – and I had noticed how the Cirrus trails were spreading across the blue as I walked through London to the railway station. There were some low-level Cumulus clouds around, whose movement showed the direction of the low wind – not always so easy to determine amongst the capricious eddies that form in the vicinity of high buildings.

Pausing, and standing with my back to the wind, I could see that the thickening high Cirrus were spreading to my right, suggesting that a region of low pressure was on its way. Did it depress me that this beautiful uninterrupted view of Cirrus would not last? On the contrary, I was curious to see if the cloud would develop in the way I had read about – keen to watch its silent performance, laid on for my pleasure.

Throughout the westward journey, I observed the warm front at the leading edge of the depression silently choreograph the clouds. Ahead of it, the Cirrus spread in anticipation, forming a milky veil, which joined, thickened and lowered into an Altostratus. Then, as if on cue, the first drops of rain gathered on the window of the carriage – tiny drops at first, but gradually developing into rivers down the pane.

By the time the train passed the boundary between cooler and warmer air at ground level, the Altostratus above had thickened and lowered still further. The steady rain fell more heavily from Nimbostratus, which hung low in the sky. Had I been watching from London, it might have taken 24 hours for this part of the low-pressure system to pass over. On the train, I was through to the warmer air in a matter of hours.

At my destination, the thinning Nimbostratus had begun to separate into broken Stratocumulus. Now in the central region of the kink, without the lifting from competing air masses, the skies were clearer. I felt sure this would pass, however, and through the afternoon Cumulus clouds began to build and one or two softened at their tops. Their glaciating upper regions indicated that they'd changed into Cumulonimbus.

By early evening, the warm hues of the low Sun were hidden from view. The sky above was bruised with dark, pregnant mounds. The clouds let rip in sudden, energetic showers. It was cooler again and I knew that the rear of the low-pressure system was now passing over – the steeper mass of cold air burrowing under the warmer, moister air on its progress east towards London. My spirits weren't dampened by the passage of the depression. I stood outside in the rain and felt the cascade of drops on my forehead – abundant, plentiful, cleansing the air with their fall. The blades of the grass twitched and shuddered in the downpour.

Enjoying the clouds is a matter of watching their progression. No cloudspotter can tell if it will rain with a quick glance at the sky. This would be akin to seeing someone's photograph, snapped at random, and knowing how they were feeling. Had it caught them mid-blink, would this mean they were sleepy? Had their face been frozen in a split-second grimace – would it mean they were in pain? No, it would mean the shot was a dud.

We need to watch a person's expression change from one moment to the next to say how they are. Likewise, beautiful Cirrus spun across the heavens will tell us little about the mood of the atmosphere. To know this, we must have the patience to watch the expressions develop.

☁

IT IS HARDLY CONTROVERSIAL to consider the clouds as harbingers of the weather. The same cannot be said of predicting earthquakes by them. But a retired Chinese chemist, now living in New York, does exactly that. Zhonghao Shou claims that the appearance of certain types of cloud is a valuable and undervalued tool in short-term earthquake prediction.

Whilst his theories are dismissed as nonsense by many seismologists, Shou is so convinced of the link between certain 'non-meteorological' cloud formations and the occurrence of major quakes that, since his retirement, he has devoted his life to poring over satellite images of cloud cover in order to make quake predictions. He claims that the appearance of 'earthquake clouds' can help him anticipate the location and magnitude of a quake with an average advance warning of thirty days.

Shou has identified five distinct types of earthquake cloud.

The most dramatic and unusual-looking are the 'line-shaped' and 'feather-shaped' ones, appearing as narrower or wider individual streaks of high cloud, much like short, straight Cirrus clouds. These

Predicting can appear very suddenly, forming in a matter of seconds,
earthquakes like the condensation trail from a rocket. 'Lantern-shaped'
from clouds earthquake clouds take the form of a line within a gap in a

pre-existing layer of high cloud. The tail of the cloud points towards the epicentre of the impending earthquake, he claims, and its length, when compared with previous sightings and their subsequent quakes, gives an indication of the likely magnitude. Shou's records suggest that a quake will take place within 103 days or less from the appearance of one of the clouds, with the average time being thirty days.

He does not claim to have a clear understanding of how quakes might affect the clouds, but offers an explanation similar to the way a volcano smokes before erupting. 'Underground water vapour, at a very high temperature and pressure, erupts to the surface from one or more crevices,' he proposes. 'It then rises up to form a cloud when it meets the cold air in the atmosphere above.' Shou suggests that the subterranean rock might develop small cracks as a result of the seismic stresses in advance of a major fault. Underground water then percolates into these crevices, becoming heated by the enormous friction. Under tremendous pressure, as it expands, the water eventually erupts to the surface as a jet of vapour, forming a cloud in the skies above. This can act as a marker – its position and orientation indicating the general location of the impending fault, its size suggesting the degree of seismic force, and so the quake's magnitude.

Without any geological training, Shou is the first to admit that the mechanism behind earthquake cloud formation needs research. As far as he is concerned, it is the accuracy of his predictions that is most important.

Since he began logging his predictions in 1994, Shou claims that around 70 per cent of his predictions have proved correct, in spite of his only having access to publicly available satellite images. Had he access to higher-resolution, continuous images, which are generally classified, he claims his success rate would be higher.

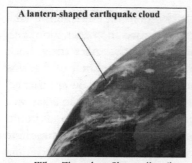
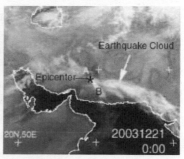

LEFT: What Zhonghao Shou calls a 'lantern-shaped' earthquake cloud. RIGHT: On 25 December 2003, Shou used this cloud to predict an earthquake of magnitude 5.5 or more. The point from which it emerged (✳) marks the epicentre of a 6.6 magnitude earthquake, which struck the following day in the Iranian city of Bam.

Those seismologists who considered Shou to be a quack began to take notice after 25 December 2003. On that day, he made an earthquake prediction on his website. Looking through images taken a few days before from the Meteosat-5 weather satellite, positioned over the Indian Ocean, Shou had identified a classic earthquake cloud along a well-known geological fault line in southeastern Iran. The images, which showed a huge trail of cloud appearing to emerge from half-way along the fault line, led Shou to predict an earthquake of a magnitude greater than 5.5 on the Richter scale in the region sometime within the next sixty days.

At 5.26am on 26 December, a quake of magnitude 6.6 struck along the fault line, with its epicentre in the ancient Iranian city of Bam – a position that corresponded almost exactly with the tip of the cloud that Shou had identified. It caused massive destruction, killing more than 26,000 people and injuring tens of thousands. Of the buildings in the 1,500-year-old Silk Route trading city, 70 per cent were flattened.

Following the remarkable success of his Bam prediction, Shou was invited in May 2004 to talk at a UN and Iranian Space Agency workshop in Tehran, which was debating the use of space technology for environmental security and disaster rehabilitation. According to Ansari Amoli, the space agency's remote sensing and disaster management expert, Shou's presentation was well received by the geologists, seismologists and meteorologists present. 'His earthquake clouds appear to be a very promising way of improving the short-term earthquake prediction, if used in conjunction with traditional methods,' says Amoli. 'But there is a need for a much better understanding of the mechanisms at work. I believe that it is an area that is worthy of serious research by earthquake specialists.'

Whether or not his methods for predicting earthquakes will begin to be accepted by the wider scientific community has yet to be seen. Some see them as far-fetched. 'Only Mr Shou thinks there is any relationship between clouds and earthquakes happening 10km below the surface of the earth,' was the comment of Dr Lucy Jones, Scientist-in-Charge of the Pasadena Field Office of the United States Geological Survey. His theories might not be as ridiculous as she makes out, however. They certainly have a long historical pedigree.

The Roman historian Pliny the Elder, drawing on the observations of Aristotle in AD77, alluded to clouds appearing in advance of earthquakes:

> *There is no doubt that earthquakes are felt by persons on shipboard, as they are struck by a sudden motion of the waves, without these being raised by any gust of wind… There is also a sign in the heavens; for, when a shock is near at hand, either in the daytime or a little after sunset, a cloud is stretched out in the clear sky, like a long thin line.*[3]

They are also described in the 32nd chapter of the *Brihat Samhita*, a Sanskrit text from the sixth century AD by the philosopher, mathematician and astronomer Varahamihira. The work, which is considered a seminal text of ancient Indian astronomy and astrology, claims that a particular type of earthquake is preceded a week before by an unusual cloud formation:

> *Its indications appearing a week before are the following: Huge clouds resembling blue lily, bees and collyrium in colour, rumbling pleasantly, and shining with flashes of lightning, will pour down slender lines of water resembling sharp clouds. An earthquake of this circle will kill those that are dependent on the seas and rivers; and it will lead to excessive rains.*[4]

The first recorded prediction of an earthquake based on the appearance of the clouds appears in the *Chronicle of Lon-De County, China*, compiled in 1623:

> *It was sunny and warm; the sky was blue and clear. Suddenly, there appeared threads of black clouds spanning the sky like a long snake. The clouds stayed for a long time, so there would be an earthquake.*[5]

Shou claims to have found records of an earthquake in Guyuan, in China's Ningxia province, on 25 October 1622, the only one of its magnitude in Western China within the 148 years between 1561 and 1709.[6]

☁

WHETHER OR NOT they can find clouds that presage earthquakes, cloudspotters certainly can see those that predict the arrival of local weather systems. It is the behaviour of the high clouds, such as Cirrus, that they should watch. When they are spreading and thickening across the blue these Cirrus seem not so much floes of angel hair, as tufted whiskers of a wise man's beard. He's a genial old fellow, who'll tell of the weather in store. But he speaks in a whisper. It is one that only those who pay attention will ever hear.

John Miles (member 1177)

Whether Cirrus are angel hair or a wise man's beard, this is presumably the comb.

NINE

CIRROCUMULUS

*The fleeting layers of rippling cloudlets,
known as mackerel skies*

At first glance, it is easy to miss the fact that the Cirrocumulus cloud is, like its lower cousins Stratocumulus and Altocumulus, composed of individual cloudlets. Being so high up – typically between 16,500 and 45,000ft in the middle latitudes – these cloud elements can appear tiny, like little grains of salt. In fact, you have to look quite carefully to see that this cloud is made up of separate elements at all. A patch of Cirrocumulus (for it tends to appear as patches, rather than cover the whole sky) often looks no more than ripples in a high, smooth layer.

But cloudspotters are not the type to just glance at the sky. On more careful inspection, they will see that the ripples consist of tiny elements. These will appear smaller than the width of a finger, held at arm's length above 30° from the horizon, for though the cloudlets themselves may be the same size as Cumulus humilis, they certainly are a long way up.

Observing the apparent size of its elements is one way of distinguishing Cirrocumulus from the lower Altocumulus (whose cloudlets are comparable to between one and three finger widths). The other way is to look at its shading. Or, rather, its lack of shading – for the higher Cirrocumulus appears as a brighter white than the mid-level Altocumulus and its elements are of a more even brightness, whereas the cloudlets of the lower Altocumulus are darker on the sides in shadow.

HOW TO SPOT
CIRROCUMULUS CLOUDS

Cirrocumulus are high patches of cloud or layers of tiny cloudlets that appear as white grains. These show no shading, even on the sides away from the Sun. These cloudlets are generally regularly spaced, and often arranged in ripples, known as the undulatus variety.

TYPICAL ALTITUDES*:
16,500–45,000ft
WHERE THEY FORM:
Worldwide.
PRECIPITATION (REACHING GROUND):
None.

CIRROCUMULUS SPECIES:

STRATIFORMIS: When it is in an extensive layer, rather than just a patch. A less common species than for other genera.

LENTICULARIS: When it is in the form of one or more independent, well-defined, almond- or lens-shaped masses, which have smooth surfaces and are much larger than the grain-like cloudlets of the other species.

CASTELLANUS: When, on careful inspection, its cloudlets have crenellated tops.

FLOCCUS: When, on careful inspection, its cloudlets are Cumulus-like, with ragged bases.

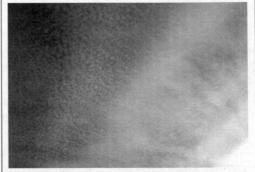
Cirrocumulus stratiformis

Cirrocumulus lacunosus undulatus

CIRROCUMULUS VARIETIES:
UNDULATUS: When its cloudlets are in a wave-like arrangement of ripples or broad undulations (or both at the same time).
LACUNOSUS: When the layer has holes fringed with cloud, like a net or honeycomb.

NOT TO BE CONFUSED WITH...
CIRRUS AND CIRROSTRATUS: which are streaks and smooth/fibrous layers of high cloud. Cirrocumulus layers, by contrast, are subdivided into many grain-like cloudlets.
ALTOCUMULUS: which is a mid-level layer of larger cloudlets. Looking above 30° from the horizon, the smaller Cirrocumulus cloudlets generally appear less than the width of one finger, held at arm's length.

* These approximate altitudes (above the surface) are for mid-latitude regions.

Cirrocumulus is the most elusive of the ten cloud genera. Indeed, when it does appear, its grains soon dissolve, representing a transitional phase between the wispy filaments of the Cirrus and the smooth, milky, layer cloud, known as the Cirrostratus. One way meteorologists identify clouds is by noting the other types it appears in conjunction with, and the presence of the easily recognised streaks of the Cirrus helps to identify the dappled delights of the Cirrocumulus.

The separate elements of the layer show that the air up at cloud level is choppy and unstable. When there is just a patch or two, this is not particularly significant of the weather in store. Occasionally, however, the formation appears as ripples over a large area of the sky – the stratiformis species of Cirrocumulus, of the undulatus variety. Easier to remember than the not-so-snappy Cirrocumulus stratiformis undulatus appellation is the name 'mackerel sky'. It is a term, most probably coined by mariners, for a cloud formation that has long been seen as a warning of approaching storms. This is particularly the case when the cloud is combined with the hook-shaped formation of Cirrus uncinus, described as 'mares' tails'.

☁

THE TERM MACKEREL SKY is sometimes also used to describe high Altocumulus clouds. But these just don't look as mackerel-like as the Cirrocumulus. That is the one whose striations most resemble the fish's distinctive stripes, its individual cloudlets appearing as the scales.

Why can large areas of rippling Cirrocumulus signify deteriorating weather? Firstly, the large area of high cloud suggests there is a lot of moisture at the top of the troposphere. In temperate regions, this can be an early indicator of an advancing depression, bringing rain. Secondly, the choppy waves in the formation indicate that the winds up at this level are strong, implying that the approaching weather system will be a strong one.

The waves of the mackerel sky are similar to those on the sea surface. Ocean waves develop as wind blows over the surface of the water, catching and amplifying any disturbances in the surface.

The wind encourages the undulations of water to rise upwards, gravity pulls them back down, and the water's wave-forms are the result of the opposing forces.

In the lofty domain of the Cirrocumulus cloud, of course, there is no clear-cut division between a mass of liquid and one of air. But a similar sort of system can develop when the cloud forms in an area of 'wind shear'. This is when air above the cloud layer is moving at a different speed and/or direction from that below. The region in between these two shearing air streams tends to undulate and – just like on the sea's surface – when the winds are high, the undulations are choppy.

☁

THE ATMOSPHERE ITSELF is an ocean – one of air, rather than water. The relationship between the atmospheric ocean and the actual one is close, and of great relevance to the formation of clouds in general.

It is easy to forget that the atmosphere starts at our feet. We are therefore much like the crustacea, crawling around on the bed of this sea of air. The cloudspotters amongst us gaze up at the birds *Cloudspotters* gliding through its currents as well as the other crustacea on *and crustacea* the move in the submarine-equivalents we call aircraft. No doubt, clouds exhibiting the evaporating trails of precipitation, called virga, which can hang below them like dangling tendrils, are the jellyfish.

The relevance of the aquatic oceans for cloud formation generally is due to the way the currents, temperatures and humidity of the air are influenced by the waters below. Covering 70 per cent of the Earth's surface, they are in fact the most important factor, after the Sun itself, in the clouds' appearance and behaviour.

For starters, 90 per cent of the moisture in our atmosphere evaporates off the sea. The remainder comes from rivers, lakes and other waterways, as well as from the leaves of plants keeping themselves cool by 'evapotranspiration' – a sort of botanical version of sweating. Oh yes, and moisture also finds its way into the atmosphere by humans sweating and sneezing, evaporation off

The ripples of the Cirrocumulus stratiformis undulatus, or 'mackerel sky'.

Chris Dolley (member 1620)

clothes on the washing line, glasses of gin and tonic after croquet, and little puppies' tongues.

It is not just that the oceans cover so much of the planet's surface that makes them so important. Water is remarkably efficient at storing heat and transporting it huge distances around the globe in the pattern of principal ocean currents. So oceans not only provide a ready supply of moisture to the atmosphere, they also have the effect of warming and cooling the air passing over them – both critical factors in cloud formation.

Tropical cyclones and hurricanes develop when atmospheric disturbances move over the sea, picking up heat and moisture over warm ocean currents. Certain atmospheric conditions have to be met for storm systems like these to get started, but once they do, the ready supply of heat and moisture from the sea surface provides the storm with its monumental energy.

Whipped into an enormous rotating system, it is an unstoppable force. Only when it passes over land – perhaps the homes of unfortunate souls in its path in Louisiana, the Caribbean or India – with all the destruction that this entails, does it finally begin to dissipate, its energy supply from the warm sea surface interrupted.

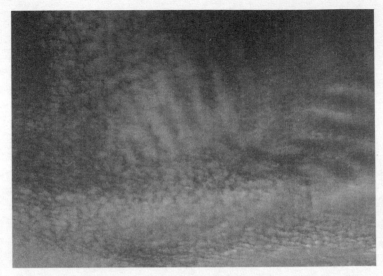

This Cirrocumulus is of the floccus species and is, in places, exhibiting billows that signify the undulatus variety.

Clouds of a much more sedate variety are associated with cold ocean currents. These can well up from below the surface on continental coasts and form large areas of low Stratus and fog. The famous summer fogs of San Francisco are a good example.

Air streams blowing in towards land pick up heat and water vapour over warm ocean currents in the Pacific. Passing across the colder water near the coast, they drop in temperature and some of their water vapour forms into cloud droplets in the process. Without having to rise to cool, the droplets form at surface level as advection fog. This makes San Francisco one of the foggiest cities in the world, though the fog is usually limited to the part of the city along the coastal strip.

Also competing for this title, however, are areas on the northeast coast of Japan. There, a similar contrast of sea surface temperatures exists. Warm, moist air blows inland from over the warm Kuroshio current out in the Pacific, only to cool as it reaches the cold Oyoshio current near to the coast. Again, the resulting drop in temperature causes extensive regions of fog or mist to sweep inland.

These fogs served as inspirational subject-matter in some styles of traditional Japanese painting. The artistic device called *kasumi*, which is Japanese for 'mist', was traditionally used as a way of giving depth and perspective to landscape scenes. They generally took the form of horizontal bands, which in the early paintings of the Heian period (around AD1000) tended to be soft and transparent, with a blue tinge. By the thirteenth century, they more commonly appeared as 'solid' patches of fog, sharply outlined in ink, and known as *suyarigasumi*.

Besides providing the landscapes with a sense of depth, the beautiful *kasumi* mists sometimes had a role in punctuating a narrative within the paintings. They signified the passage of time between different scenes in an image. Never have 'the mists of time' been more literally expressed in art.

☁

THE STRATIFORMIS UNDULATUS of mackerel sky fame is not the only formation of Cirrocumulus. Besides stratiformis, where the layer stretches over a large area of the sky, there are three other species, depending on the appearance of the cloudlets.

Castellanus is when the elements have turret-like tops, which rise from flat bases. With the individual elements being so high above, however, it is harder to identify these turrets than with the castellanus species of the lower clouds, such as Altocumulus and Stratocumulus. The same can be said of floccus, when the clumps have uneven bases as well as uneven tops. Both are indications of vigorous growth of the individual cloudlets, which happens when the air up at the cloud level is what is described as 'unstable'.

By contrast, the lenticularis species occurs when the air is 'stable' and it is quite different in appearance from the others. This is when a relatively large region of cloud is formed into a lens shape. It is a higher version of the UFO lenticularis formations that appear at the lower levels. Here, the rule which states that Cirrocumulus cloud clumps look smaller than the width of a finger breaks down – lenticularis elements can be much larger than this. It is worth dwelling for a moment more on the Cirrocumulus

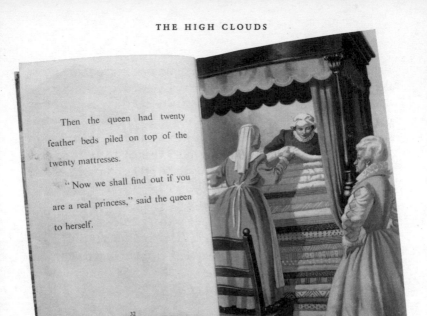

Then the queen had twenty feather beds piled on top of the twenty mattresses.

" Now we shall find out if you are a real princess," said the queen to herself.

32

The Hans Christian Andersen fairytale, *The Princess and the Pea*, shows why a 'stable' atmosphere is conducive to the formation of Cirrocumulus lenticularis.

lenticularis cloud, as it helps to explain the important concept of atmospheric stability.

Like the equivalent lower formations, it appears when air is forced over mountain ranges, and develops a wave-like motion in the lee of the peaks, with the lens- or almond-shaped clouds appearing at the crests of the waves. It might seem strange that air streams flowing over obstructions down at ground level – even obstructions as high as mountains – can produce waves that form clouds at an altitude of 26,000ft (5 miles) or more. In fact, it doesn't happen very often, and it depends on all the air between the ground and the cloud being stable.

Whether a region of air is described as stable or unstable depends on how the temperature changes with height. The distinction between the two is quite involved (a region of the atmosphere is actually said to be stable or unstable relative to a 'bubble' of air at a certain temperature and humidity). At its most simple, the air is more likely to be unstable when it becomes suddenly colder with altitude, whilst a more gradual cooling is described as stable.

This temperature profile has an important role in the formation of clouds. In the case of Cirrocumulus lenticularis, for example, the stability of the air above the mountain range determines how 'springy' it is, which is a critical factor in whether the waves that develop in the lee of the range reach the air much higher above.

A stream of air, forced to rise as it passes over a mountain, expands and drops in temperature, as any rising air does. But if the part of the atmosphere just above the air stream is a lot colder, the rising air can, in spite of having cooled, still be warmer than it. The rising air will then tend to float upwards and the surrounding air sink around it. The atmosphere above is unstable relative to the air stream, and absorbs the crest of the wave, without the air above being pushed upwards.

If, by contrast, the atmosphere above grows colder gradually with altitude, the air stream – rising and cooling as it flows over the mountain – might end up at around the same temperature. The atmosphere above is stable relative to the air stream, which doesn't float through it, but pushes the air up with it.

I suppose it is like the Hans Christian Andersen fairytale, *The Princess and the Pea*. This tells how during a violent storm a princess arrives, sodden, at the gates of a castle, where the old king and queen are keen to have their son married off. She seems to have 'daughter-in-law' written all over her, but they want to be sure that she is a real princess. They offer her lodging *Cirrocumulus lenticularis and fairytales* for the night and, with a logic that is the sole preserve of prospective mother-in-laws, the old queen decides to set a secret test by preparing a bed with a pea at its base, upon which she places twenty mattresses and twenty eiderdowns. The princess sleeps terribly and they conclude that she is surely sensitive enough to be the genuine article. The prince promptly marries her and they live happily ever… blah, blah, blah.

Unstable layers of air above will, like very soft mattresses, absorb the rising crests of the air stream flowing over a mountain. No matter how pronounced they are, these will not make the air much higher in the atmosphere rise. Stable layers, however, will all be pushed upwards with the wave and the atmosphere many miles above will 'feel' the pea-like crest and rise up slightly with it. If the air all the way up there is moist enough, a Cirrocumulus lenticularis can form at the rise.

Clearly, this makes the lenticularis a genuine princess. It means, no doubt, that the Cumulonimbus king will consent for her to marry his Cumulus son. I don't know which type of cloud the old queen is, but I'm sure they all live happily ever after nonetheless.

☁

THE DIFFERENT SPECIES of Cirrocumulus can exhibit either (or neither) of the two possible varieties for this genus, namely lacunosus and undulatus. The appearances of each correspond with their lower cloud equivalents.

Lacunosus is when the Cirrocumulus is a lattice of cloud around distinct holes. Being so high up, this loose honeycomb formation is finer than the lacunosus varieties of Altocumulus and Stratocumulus.

Cirrocumulus undulatus is when the cloudlets gather into waves, which appear as bands. On occasions, there can be two wave-forms superimposed upon each other – the cloudlets aligning in broad undulations as well as smaller ripples. Such a formation is analogous to large ocean waves having smaller waves along their surface. In both cases, the superimposed wave-forms need not necessarily travel in the same direction.

More often, however, the undulatus variety has just one wave-form in evidence, and this is usually the case with the Cirrocumulus stratiformis undulatus of the mackerel sky.

No doubt, all this talk of the Cirrocumulus species and varieties will have thrown up many questions for the keener cloudspotter. The most nagging, of course, will be: what kind of mackerel is it that a mackerel sky looks like? Is it the king mackerel, or the Spanish mackerel, or is it just the common old Atlantic mackerel? Deciding that it wouldn't do to leave such an important question unanswered, I went on a mission to find out.

The fine honeycomb formation of Cirrocumulus lacunosus.

WAKING AT 5AM on a clear August morning, I took the first tube across town to visit the Billingsgate Fish Market on the Isle of Dogs, in London's East End. The market boasts the largest selection of fish in the United Kingdom, so I figured it would be as good a place as any to compare the markings of mackerel to Cirrocumulus clouds. Of course, I had no idea whether the particular cloud formation would be in evidence that morning. Nor did I know if the traders would be kind enough to lend me their fish so that I could hold them up and compare.

Emerging from the underground station, amongst the high-rise offices of Canary Wharf, I was pleased to see that there were indeed some patches of Cirrocumulus above, amongst dramatic Cirrus fallstreaks. Billingsgate is housed in a large market hall, however, so even if this Cirrocumulus were to develop into the stratiformis undulatus species I would still have to depend on memory alone. With a mental image of the cloud in mind, I swept into the hall and wove my way through the hubbub of mongers, porters and restaurateurs. I was a man on a mission – a mackerel mission.

The easiest to find was the Atlantic mackerel. It is the most common member of the family to appear on British shores. I gravitated towards a bunch on ice in a polystyrene crate and stared closely at the iridescent silver and dark grey stripes that glistened down the fish's backs.

'Can I help you, mate?' asked the fishmonger – his white overall smeared with innards.

'I'm just browsing, thanks,' I replied, resisting the urge to add, 'but I'm disappointed to see that your mackerel look nothing like Cirrocumulus stratiformis undulatus.'

You see, the markings on the Atlantic mackerel were much too strongly delineated. Cirrocumulus, like all the high clouds, have softer edges than the lower clouds – this is the result of them being composed in part, if not entirely, of ice crystals. These mackerel had sharp distinctions between their pale and dark stripes.

But this wasn't the only problem. Though the silvery scales of the fish's pale stripes certainly looked the part, the regions inter-spersing them were much too dark to suggest the sky. They looked almost black.

I tried to imagine a moonlit Cirrocumulus stratiformis undulatus – one with the black of night showing between its bright bands of cloudlets – but it just wasn't working. The waves of Cirrocumulus cloudlets have paler, softer tones against the blue of the heavens. Clearly my fishy quest would not be ending at the Atlantic mackerel.

☁

'I AIN'T GOT NO SPANISH mackerel, mate,' said the stallholder, when I made enquiries about the next recruit in my fish identity parade. 'We don't get them these days,' he added, wistfully. 'I 'aven't seen many Spanish 'ere for years.'

Damn. The market with the largest selection of fish in the United Kingdom didn't even have Spanish mackerel. I feared that my crack-of-dawn outing was going to prove pointless. But then the fishmonger gave me a hot tip. If I could find someone selling young king mackerel this would do the job. 'Adolescent kings,' he

The markings of the Atlantic mackerel. Sadly for this fish, they are too sharply defined to be responsible for the term 'mackerel sky'.

whispered furtively, looking from side to side, 'look a lot like mature Spanish mackerel.'

Actually, he didn't whisper this furtively – he just said it normally.

King mackerel, or kingfish, was one of the ones I was after anyway. Now, if I could find a young one, as well as the adult, it could act as an identity parade stand-in for the absent Spanish.

I picked my way past hake, bass, bream and turbot. Past dogfish, monkfish, conger eels and lobsters. The creatures of the deep were having a mesmerising effect on me. Red snapper, grey mullet, whiting… and finally I found what I was after – some young king mackerel, on a stall at the side of the hall, near to the crabsticks.

The young king was twice as large as the adult Atlantic mackerel, and had entirely different markings. Its belly was a smooth silver colour, merging to pale blue up its sides. And across the blue were a series of round yellow spots.

Wait a minute, this looked even less like the cloud. There was nothing 'mackerel sky' about these markings – the spots were too far apart to suggest Cirrocumulus, and there were none of the all-important ripples. If this fish was indicative of the colouring of a Spanish mackerel, then it would be laughed off my identity parade before it had even flapped its way to the line-up.

Just a few stalls on, however, was the impressive sight of an adult king mackerel – this was much more like it. It was a lot bigger

– around three feet long – and had lost its yellow spots. Running through the iridescent silvery blue of its sides were pale ripples of white and silver. Bingo!

Here, on the back of this impressive catch, were the undulated markings of the Cirrocumulus's mackerel sky – beautiful, curving alleys of silver scales, separated by the pale blue of the sky. At £8 per kilo, I had finally found the very fish that gives its name to the mackerel sky.

What delight, to have cleared up this vexing matter. I strolled back through the hall, with the satisfied air you'd expect from *Mission* a world expert in the specialised field of fish/cloud *accomplished* comparison. Of course, I mused to myself, Spanish mackerel may not have looked like a true mackerel sky, but its yellow spots *were* rather reminiscent of widely spaced Altocumulus, lit by the amber hues of sunrise…

And then, I stopped in my tracks. I was staring at a large fat carp, over by the stall with the Alaskan smoked salmon. It stared back at me, unblinking, as dead fish do.

It couldn't be… The fish's scales, broad for its body size and ranging from a muddy yellow on the carp's belly to a deep bronze at its spine, had something very cloud-like about them. Each scale was amber in the middle, growing darker brown towards the edges. I recognise this sky, I thought… Come on, come on – you're the world expert, for God's sake – what sky is on the carp…?

Of course! The Altocumulus stratiformis perlucidus! How could I have hesitated? This cloud was like an old friend to me – I just didn't recognise it out of context like this.

A lower cloud than the Cirrocumulus, Altocumulus's cloudlets appear larger – in keeping with the carp's large scales – and, in the light of a low Sun, they are darker on the sides in shadow – just as the scales are darker at the edges. These scales could never have been Cirrocumulus, whose cloudlets – as we all know – exhibit no shading. They were Altocumulus stratiformis (a layer covering a large area of the sky) perlucidus (with small gaps between the cloudlets). It can only be a matter of time, I decided, before this formation becomes known as a 'carp sky'.

Being a freshwater fish, which hangs around in the murky

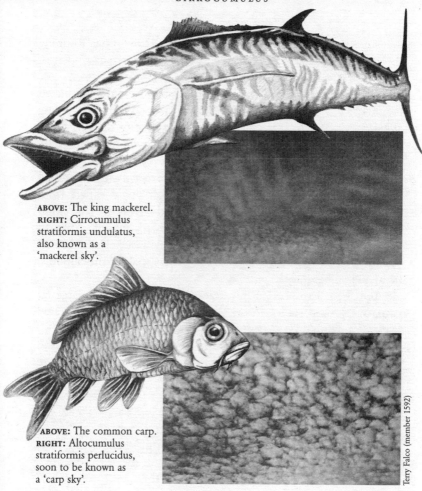

ABOVE: The king mackerel.
RIGHT: Cirrocumulus stratiformis undulatus, also known as a 'mackerel sky'.

ABOVE: The common carp.
RIGHT: Altocumulus stratiformis perlucidus, soon to be known as a 'carp sky'.

Terry Falco (member 1592)

depths of turbid lakes, the common carp couldn't be more different from a deep-sea game fish like the mighty king mackerel. How appropriate, then, that a carp sky foretells little more than the approach of some light rain. This is no warning to hoary old mariners to stow the mainsail and batten down the hatches for a furious Atlantic storm.

No, a carp sky is more a reminder for the drowsy angler to unpack his Barbour in a few hours, as there could be a spot of drizzle before tea.

TEN

CIRROSTRATUS

*The high milky veils,
which most people barely notice*

S eventeen hundred years ago, the Cirrostratus cloud changed the course of human history. It was responsible for a chain of events that resulted in Christianity becoming the dominant religion of the Roman Empire.

At least, that is how we cloudspotters see it.[1]

On 28 October, AD312, Emperor Flavius Valerius Constantinus – also known as Constantine the Great – defeated his rival and brother-in-law, Emperor Augustus Maxentius, at the Battle of Milvian Bridge, just north of Rome. The two emperors were vying for control of the western regions of the Roman Empire. With only 50,000 men to Maxentius's 75,000, Constantine emerged the victor and went on to become the most important emperor of late antiquity. Not only did he spread Roman influence into the Middle East by establishing a 'New Rome' at Byzantium (later Constantinople, now Istanbul), but he legalised and supported the Christian religion, which had previously been outlawed in the Roman Empire.

Constantine's victory at Milvian Bridge was undoubtedly a decisive moment in world history, and – if some historians of the time are to be believed – it was the result of a miraculous sign in the sky, revealed to Constantine on the eve of the battle.

Some 25 years later, Bishop Eusebius of Caesarea wrote an account of the legend surrounding this vision in his *Life of*

HOW TO SPOT
CIRROSTRATUS CLOUDS

Cirrostratus are largely transparent, milky veils of high cloud that look either smooth or fibrous. They tend to cover large areas of the sky, extending over many thousands of square miles, but are often so subtle as to be missed. They do, however, sometimes produce the white or coloured rings, spots or arcs of light around the Sun or Moon that are known as 'halo phenomena'.

TYPICAL ALTITUDES*:
16,500–30,000ft
WHERE THEY FORM:
Worldwide.
PRECIPITATION:
None.

HALO PHENOMENA:

Cirrostratus causing a
'22° Halo' around the Moon

Sundog, or mock sun

Cirrostratus fibratus causing
a 'sundog' at the same
elevation as the Sun

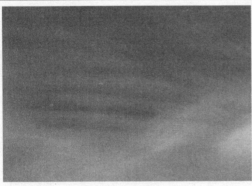

Cirrostratus undulatus

CIRROSTRATUS SPECIES:
FIBRATUS: When the cloud veil has a fine fibrous or striated appearance.
NEBULOSUS: When it shows no variation in tone.

CIRROSTRATUS VARIETIES:
UNDULATUS: When the veil has a wave-like appearance.
DUPLICATUS: When there is more than one layer, at different altitudes. This is generally only visible when, by the light of a low Sun, the higher layer is lit up when the lower is in shadow, or when shearing winds cause the striations of each layer to differ.

NOT TO BE CONFUSED WITH...
ALTOSTRATUS: which is a mid-level, generally thicker, layer cloud. Besides being thinner, the ice crystals of the Cirrostratus can sometimes produce halo phenomena around the Sun or Moon. These are far less common in Altostratus, which will generally only produce a corona (a white or coloured disc of light).
CIRRUS OR CIRROCUMULUS: which are streaks and grained/rippled layers of high cloud. Cirrostratus, which often appears in conjunction with them, is a more continuous and diffuse layer.

* These approximate altitudes (above the surface) are for mid-latitude regions.

Top right: Peg Zenko (member 1527)

Constantine (*c.*337–339). He claimed that, whilst marching towards Rome on the day before the battle, Constantine and his army saw a cross of light in the sky, above which was written '*hoc signo victor eris*' – 'by this sign you shall be victor'.

That night, according to Eusebius, Christ appeared to Constantine in a dream 'and commanded him to make a likeness of that sign which he had seen in the heavens, and to use it as a safeguard in all engagements with his enemies'. So he duly instructed standards to be produced with the sign on them. His army marched to victory behind this symbol, which became known as the *labarum*.

The symbol subsequently appeared on numerous Roman coins associated with the decisive battle and, with the rise of Christianity, it later became seen as a symbol of that faith. It usually appeared as a diagonal cross – like an 'X' – with a vertical line rising from its centre and the bowl of a 'P' at its top.

Eusebius's account of Constantine's vision doesn't completely tally with those of other historians of the time, but he does say in his book that Constantine told it to him, late in life, and 'confirmed it with oaths, when I was deemed worthy of his acquaintance and company'. So, you can't argue with that, can you?

Various arcs, lines and patches of light can appear in the sky when the Sun's rays are refracted, as they pass through the ice crystals of a Cirrostratus cloud. These optical effects, known collectively as 'halo phenomena', might just account for the elements of the *labarum* symbols on Roman coins commemorating Constantine's victory. *Cirrostratus and the rise of Christianity*

I should point out, however, that at the time of writing no one has recorded any of the cloud's halo phenomena that are in the shape of '*hoc signo victor eris*'.

THE CIRROSTRATUS is a delicate-looking layer of ice crystals, which tends to form at altitudes between 20,000 and 42,000ft. It generally appears as a pale, milky lightening of the sky and often forms from the spreading and joining of Cirrus – a cloud it is often

Cirrostratus fibratus duplicatus. You can see there is more than one layer (duplicatus), since the winds have given the striations different orientations.

seen in conjunction with. At times, the layer can be so thin as to be barely noticeable – just the faintest opalescence over the blue above. At others, it is a more distinct, milky white, though never thick enough to block the Sun completely.

There are just two species of Cirrostratus: fibratus and nebulosus. Cirrostratus fibratus, like the corresponding Cirrus species, has a striated texture – like fibres of silk – and is therefore easier to spot than the smooth, featureless Cirrostratus nebulosus.

The varieties of Cirrostratus number the same: just duplicatus and undulatus. As in other cloud genera, the former is when the cloud is in more than one layer, these being at different altitudes. This is as good as impossible to see in the full light of day, since one layer above another looks no different from a single, thicker one. With a low Sun, however, the angle of the light can make the layers more readily distinguishable. For moments around sunrise and sunset, the higher layer of Cirrostratus duplicatus can be lit, whilst the lower one is in shadow.

Undulatus is when the layer has a rippling underside. Even when the Sun is low, the cloud is not generally dense enough for the undulations to be cast in strong shadow, as those of Altostratus undulatus would be, but the tones from the glancing sunlight can be enough for them to be visible. Undulatus is easier to notice when the gaps between wave crests are almost transparent.

The most likely confusion in spotting Cirrostratus is to mistake it for the lower, mid-level Altostratus cloud. Generally speaking, Cirrostratus blocks the Sun much less than Altostratus, and so sunlight through Cirrostratus is almost always strong enough to cast shadows on the ground, whereas sunlight through Altostratus is usually so diffuse as to cast none.

THE MOST CONCLUSIVE WAY of identifying a pale layer of cloud as Cirrostratus is through its halo phenomena. Although this cloud certainly doesn't always produce these arcs, rings and spots of light, their presence is evidence enough to identify a layer cloud as Cirrostratus. Cloudspotters would be wise to acquaint themselves with these beautiful displays, and look to the various regions of the sky in which they can appear whenever the heavens pale to a milky opal.

The first time I noticed one, it looked as if the cloud was smiling at me. High above the dazzling Sun, which was barely weakened by the Cirrostratus's thin layer of ice crystals, was an arc of colours. It was like a segment of a circle, centred on the point of the sky directly above me, and looked like a miniature, upturned rainbow. The colours were brighter than those of a rainbow, ranging from blue at the top lip of the smile, through green and yellow, to red at the bottom lip. No one could have mistaken it for a common or garden rainbow. It was in completely the wrong part of the sky to be one, for rainbows are only ever visible when the Sun's behind you. This arc of colours was high in the sky above the Sun. I decided, then and there, to call this Mona Lisa of the vapours a 'cloud smile'.

I was disappointed to learn, not long after, that this beautiful arc of light already had a name. You can imagine how upset I was to discover that it is actually called a 'circumzenithal arc', or CZA. I know that a cloud smile sounds a bit flower-power, but surely they could have come up with a more evocative name than CZA?

The cloud smile – otherwise known as a CZA – appears when sunlight passes through a layer of Cirrostratus consisting of ice crystals that are clear and in the shape of tiny hexagonal plates, a few tenths of a millimetre across. The cloud crystals have a large number of possible shapes and sizes, the form they take being determined by the temperature and humidity of the conditions in which they grow. Given that it depends on this particular shape of crystal, it is no wonder that a Cirrostratus cloud doesn't grace us with a smile on its every appearance. It is even less surprising when you learn that for the CZA to appear, these tiny hexagonal plates have to be in a certain orientation – they all have to have their broad sides more or less horizontal. Clearly, this is a rather pedantic cloud – only happy when its crystals are just so. Luckily, plate-shaped crystals do have a habit of orientating themselves horizontally. In the absence of too much turbulence at cloud level, they can fall gently through the air like leaves on an autumn day.

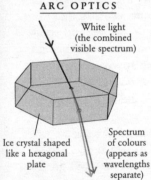

CIRCUMZENITHAL ARC OPTICS

White light (the combined visible spectrum)

Ice crystal shaped like a hexagonal plate

Spectrum of colours (appears as wavelengths separate)

The circumzenithal arc is explained by sunlight passing through the top and side of hexagonal plate-shaped crystals.

When the crystals are just right, those in certain regions of the sky will act as tiny prisms that point the sunlight into the cloudspotter's eyes. The light changes direction as it passes in through the top of the plate and out through one of its sides (these faces being at 90° to each other). Each of the wavelengths of the sunlight changes direction to a slightly different degree, which is why the light of the CZA is dispersed into a spectrum of wavelengths, appearing as rainbow colours.

That first time I saw a cloud smile was on a London street, and no one else seemed to be paying the slightest bit of notice to the sky. I was transfixed, of course, but the passers-by all had other things on their minds. I felt as if I was the only one watching this particular smile. In fact, looking back on it, I can say that

I most definitely was the only one. Even if others *had* been staring up, they would not have seen the same CZA as me.

As sunlight passed through the countless crystals up in the cloud, it was being scattered in all directions. But it was only those crystals that sparkled light directly to my eyes that created the light effect for me. Some of them flashed a little red-looking light towards me, others a little blue.

Say some of the bustling Londoners had turned out to be cloudspotters in disguise. Had they dropped their shopping and stood beside me to look up at the coloured arc too, the array of crystals sparkling directly into their eyes would have been different ones from mine. They would have seen a different circumzenithal arc. We would each have seen our own smiles.

⌒

A CIRCUMZENITHAL ARC will tend to appear on around thirteen days per year to any cloudspotter in Europe, according to the 'German Halo Research Group',[2] which has averaged the observations of its members around the continent. This makes the CZA only the fifth most common of all the halo phenomena.

The bright patch in the centre of the image is a sundog, or parhelia, caused by the sunlight being diffracted by the ice crystals of the Cirrostratus cloud.

Much more frequent are 'sundogs', also known as 'parhelia' or 'mock suns'. They are not arcs, but points of light that appear on either side of the Sun. They form at the same elevation as it and 22° away from it on either side. This is about the span from thumb to little finger of an outstretched hand. The points of light are usually coloured red on the side towards the Sun, and yellow and white away from it. Sundogs do not always appear on both sides of the Sun – when it is shining through a cloud layer that is not particularly extensive, you may only see one.

Sundogs can appear at the same time as the cloud smile, for the same hexagonal plate crystals, falling horizontally, cause them. In their case, however, the sunlight passes in through one of the side faces and out through another, which is at 60° to it.

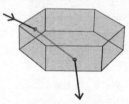

SUNDOGS OPTICS

Sundogs are explained by sunlight passing through the sides (at 60° to each other) of hexagonal plate crystals.

The German Halo Research Group tells us that sundogs are surprisingly common. They occur on around 70 or so days a year in Europe, though they do form more in the winter than the summer. Given that they are so common, it seems strange that so few people claim ever to have seen them.

Jack Borden is a retired TV news reporter turned sky fanatic, who runs an organisation in the US called For Spacious Skies.[3] Its aim is to develop people's awareness and appreciation of the sky. Over the twenty years or so that he has been running the organisation, Jack has made a point of asking people if they have ever seen a sundog. 'I got the idea of establishing it as something of a litmus test for sky-awareness,' he says. 'I'd simply ask whichever groups I was talking to, who amongst them had ever seen sundogs. Most had no idea what I was talking about, so I showed them photos of the phenomenon.' Jack estimates that only five out of every hundred people have ever seen sundogs, and out of those, two or three will have only seen them once. This doesn't sound many for a phenomenon that occurs more than once a week. Clearly, Jack has his work cut out.

Orly Doron (member 1622)

The 22° halo of the Cirrostratus cloud.

CIRROSTRATUS ARE NOT the only clouds that produce halo phenomena. These can also appear in patches of Cirrus, in the ice-crystal anvils of Cumulonimbus, and the virga ice-crystal trails of precipitation that can fall from high clouds like Cirrocumulus. The Cirrostratus stands out, however, for unlike the others it is often spread evenly over a large area of the sky. This means that the light effects can appear in more complete forms and with greater purity.

The most common of all the Cirrostratus halo phenomena is – wait for another catchy name – the '22° halo'. It is even more common than the sundogs – appearing around a hundred times a year above anywhere in Europe. Along with the much rarer '46° halo' (only four appearances a year), it is the result of the cloud's ice crystals being hexagonal columns, rather than plates. Whilst all the light effects that result from ice crystals behaving as tiny prisms are called halo phenomena, these rings around the Sun are the literal halos. The smaller 22° one is also commonly spotted at night, around a bright Moon.

22° AND 46° HALO
OPTICS

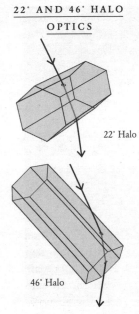

22° Halo

46° Halo

The 22° and 46° halos can be explained by light passing through hexagonal column-shaped crystals.

In the day, it appears as a complete or broken ring around the Sun, the gap between it and the Sun being a little wider than the span, thumb to little finger, of an outspread hand at arm's length. The sky is darker within, compared with outside, the ring, which has a sharp inner edge and a feathered, graded outer. It is often just white but it can display colours when sharply defined – these being red at the inner edge, through yellow, green and white, running to blue.

The much rarer, and considerably larger, 46° halo is much less bright than the smaller halo. When it does appear, the gap between the Sun and the inner edge of the halo is wider than the span of two outspread hands, held up thumb-to-thumb in a sort of prehistoric cloud-worshipping gesture. (Sadly, no halo phenomena can be measured by dropping to your knees, bowing your head and chanting.)

The crystals that form both halos are hexagonal columns like very short unsharpened pencils. The halos can be explained by the fact that these pencils fall in random orientations – not aligned, like the crystals of the CZA and sundogs. While they *can* be explained in this way, no one is sure why the crystals wouldn't tend to line up owing to the air resistance as they fall. In fact, the most common of the halo phenomena, the 22° halo, like the 46° one, is the least understood. As with all the halo phenomena, the crystals must be 'optically pure' – the pencils must be made of clear ice.

Both of these halos form from the same ice crystals, but they depend on the light passing through in different manners. The smaller halo forms when the sunlight passes in through one of the column's sides and out another one, with the faces being at 60° to each other. The larger one forms when the sunlight passes in through the side and out through the end of the column.

The tiny ice pencils often don't form with nice flat ends – these commonly look as if they've had little cones dug out of them. This is why the 46° halos appear less often. It is not because the pencils have tiny erasers on their ends.

☁

CZAs, SUNDOGS, 22° AND 46° halos are just some of the many different halo phenomena of the Cirrostratus cloud. In fact, it has quite a few more up its icy sleeve, their names sounding almost exotic: the 'upper tangent arc', 'parhelic circle', 'anthelion', '120° parhelion', 'Tricker arc', 'Parry arc', 'Hastings arc', 'Wegener arc' and 'circumhorizon arc'.

The degree of rarity of each of these phenomena is determined by how commonly the particular crystal size, orientation and optical purity, which it depends upon, occur in unison. Some are also greatly dependent on certain elevations of the Sun. Certain halo phenomena are extremely rare and only tend to appear in polar regions, where large ice crystals grow slowly to form more regular shapes with a greater optical purity. A few are so rare that they are barely more than suppositions – anticipated by computer models, which plot the passage of light through hypothetical crystals. The 'Kern halo' is one that has not yet even been photographed.

At the poles, the phenomena can commonly form as a result not of clouds, but of low-level precipitating ice crystals, called diamond dust. This is similar to a frozen fog, though the crystals actually fall, like the gentlest snow. They don't originate in cloud cover, but by growing around ground level, when temperatures are below –20°C. The beautiful glittering diamond dust of the poles can form the most dramatic and extensive halo phenomena of anywhere in the world. On one expedition to the South Pole in 1999, scientists were treated to a particularly outstanding display, in which they identified at least 24 different halo phenomena occurring at the same moment.

Thankfully, cloudspotters don't have to hang around the poles to spot the common 'sun pillar'. It looks like a broad line of light extending above (and sometimes below) a Sun low on the horizon.

Though generally classed as a halo phenomenon, this is rather different from the others, for it isn't caused by light passing *through* the cloud crystals, but by merely reflecting off their faces, when they are oscillating around a horizontal orientation. Since any flat crystal can reflect the light, this effect is not dependent on optically pure prisms or ones with clean, sharp edges. A strong, tall pillar does depend, however, on the crystals wobbling as they fall.

☁

SINCE THE SUN SHINES so brightly through thin Cirrostratus, it is perhaps understandable that people often don't notice halo effects – they know better than to look into the Sun's blinding rays.

Every book I have read about optical effects in the sky warns readers not to stare directly at the Sun. And yet, I've never come across any angry halo spotters sporting white sticks. It would seem *A disclaimer* to make sense for us to have evolved with an aversion to finding ourselves in that predicament. Nevertheless, to avoid costly lawsuits, here is the nannyish advice once again: cloudspotters should mask the Sun with their hand or stand in careful alignment with a tree when looking for halo phenomena, or else they won't be in a position to spot anything – let alone clouds.

This is not a danger, of course, when the Cirrostratus cloud's halo phenomena form in moonlight, which they can do. But, since the light from the Moon is much weaker than that of the Sun, they are generally only noticeable when the Moon is at its fullest. Even then, the light is usually too weak for our eyes to distinguish any of the colours.

☁

THOUGH BROADCAST on an enormous scale around the zenith, the spectral arc of the Cirrostratus's smile carries a message only a cloudspotter will understand. It is a whisper of the cloud's composition at the most intimate scale, telling the cloudspotter the very shape and orientation of the cloud's tiny crystals as they gently cascade through the icy air, five miles above.

But the crystals of the high clouds, like the Cirrostratus, are not always the hexagonal plates or columns that produce light effects. On the many occasions when a Cirrostratus draws its milky veil across the sky without the fanfare of optics, it can be composed of any of a myriad of other crystal formations. These may not have the right shapes, sizes or clarity to behave as perfect little prisms, but they are how the ice clouds express their individuality. They are microscopic fashion statements, which follow the trends of the upper troposphere, and each season boasts a glorious collection of styles, determined by the particular temperature and humidity of the air in which they grow.

The classics of ice crystal fashion are, without doubt, the 'stellar dendrites'. These usually comprise six identical arms, arranged in a plane – each one bristling with intricate, fractal sub-branches. They feature prominently on the glossy pages of photography books about snow. Some clouds like to be a little different from the others, sporting the classic-with-a-twist look of twelve-branch dendrites – the environmental conditions that give rise to these are still a mystery. Bolder clouds have been known to sport stellar dendrites as large as 5mm across.

Less ornate, but with equally pleasing symmetry, are the 'sectored plates' – flat slivers of ice, with the same six arms as the dendrites but a chunkier, more angular style. These look as if they've been punched from the thinnest sheets of ice – each crystal cut with its own unique die.

When crystals take the form of columns, they need not be as straightforward as the perfect little prisms that produce clear halo phenomena. 'Hollow columns' have hexagonally arranged sides but look as if tiny cones have been bored from each end – as if they have been tooled in the finest celestial ateliers, using the most intricate countersink drills.

Sometimes, columns grow very long and thin, when they are known as 'ice needles'. Cascading through the upper air, they might as well have fallen from the workbench of a seamstress in the sky. At other times, they can grow first in one direction and then, as they fall into a region of different temperature and humidity, in another one. In this way, they develop a narrow central shaft with

Dr Kenneth G. Libbrecht (member 1528)

The glittering secrets of ice-cloud fashion: from the finest 'ice needles' to the spool-shaped 'capped columns'; from the classic *haute couture* of 'stellar dendrites' to the more street look of 'rime' deposits.

a wide hexagonal plate at the end, known as a 'capped column'. It is not uncommon for a plate to develop at both ends of the central shaft, so that it now looks as if the clumsy seamstress has dropped her spools, as well as her needles.

The speed at which a cloud's crystals grow depends on the temperature and humidity of the surrounding air, and appears to be the crucial factor in determining their shape. The faster they grow, the more complex and intricate their forms.

As anyone in fashion knows, the secret of style is in combinations. As crystals fall through very different regions of air, they can take on combined forms, such as plates, columns or stellar dendrites with additional dendrite branches sprouting from them at strange angles.

When ice finds its way to the ground and falls as snow, it will have passed through many different temperatures and humidities on the way down, and will frequently have played its part in several cloud formations as it did so. No wonder, then, that snow is often in the form of a tangle of individual crystals, generally referred to as 'snowflakes'.

The shape of crystals becomes less regular as they fall through clouds of liquid droplets, which tend to freeze on to them as 'rime', roughening their sides or making them fur up, like the element of a kettle. It is more of a street look, compared with the timeless *haute couture* of elevated, pure crystals.

Despite the stunning range of crystal forms, there is one theme that keeps appearing season after season – the number six. The arms of the stellar dendrites and the sectored plates, the edges of the hexagonal plates, the sides of the columns… when it comes to ice crystals, six, rather than three, is the magic number. This is due to the shape of water molecules, which determines that as they join to form crystals, they do so in a lattice formation of hexagons – a molecular honeycomb.

THE HALO PHENOMENA of the ice-particle clouds are not the only optical effects from the interplay of sunlight and airborne water particles. There are a whole range of other optical phenomena, associated with clouds other than the high ones like the Cirrostratus. You can divide them into three main groups:

1) Those, such as 'crepuscular rays', which are shafts of sunlit air separated by cloud shadows, made visible by the scattering of light by airborne particles.

2) Those, such as rainbows, and the less familiar 'cloudbows', 'fogbows' and 'glories', which can be explained by sunlight being reflected and scattered back from drops of water, usually rain or droplets, such as those in cloud and fog.

3) Those, such as coronae and 'irisation', which can be explained by the sunlight passing around very small droplets or ice particles between the observer and the Sun or Moon.

The crepuscular rays of the first group are the dramatic shafts of sunlight that can seem to radiate from behind a dense, puffy Cumulus cloud shielding the Sun. Sometimes they shine down through a hole in a thick cloud layer – of Stratocumulus, perhaps – like the gaze of some unseen god, made visible by the vapours.

They are nothing more than shafts of light, made visible by the scattering effects of tiny water droplets (and other particles) in the air. Though not so plentiful that they appear as cloud, such droplets can still scatter the light enough for the sun-lit shafts to *The rays* stand out from the air in shadow. The effect is just like the *with the silly* way beams of sunlight become visible in the incense-laden *name* air of a Catholic church, or in the smoky atmosphere of that other place of afternoon worship – the local pub. Although the direct rays of sunlight passing through the atmosphere are as good as parallel to each other, the perspective as they approach us makes them appear to diverge.

Even when you know the explanation for crepuscular rays, it is hard not to think of them as somehow divine in nature. In Hellenistic and Roman art, emperors were often depicted with a crown of rays, known as a 'radiate'. This signified their association with the sun gods, Helios and Sol, and was also used as a sign of posthumous divinity. With the rise of Christianity, the symbol was

A Baroque fresco in the Basilica of the Crucifix, Amalfi, Italy, shows crepuscular rays being used to signify that the image of a dove represents the Holy Spirit.

Adeline Ramage (member 147)

Crepuscular rays appear between cloud shadows when the sunlight is scattered by particles and droplets in the atmosphere.

dropped in place of a circular halo, called a 'nimbus'. The rays of the radiate were felt to be too pagan in association.

The circular nimbus was used in Christian art to signify a subject's spirituality until the early Renaissance, when painting discs over people's heads began to jar with the general move towards naturalism.

In Italian art of the late sixteenth century – most notably that of Tintoretto – bursting rays of sunlight began to appear again. Being a phenomenon often seen in the crepuscular rays of Nature, a radiate nimbus was readopted as a more naturalistic way of signifying a character's divine status. It became the standard symbol throughout the Baroque period and beyond. Clearly, it is hard to resist the spiritual connotations of crepuscular rays. They emanate from behind the clouds as if originating at some unseen point in the heavens – no doubt, the place where gods hang out.

I was four years old when I first noticed crepuscular rays. It was as I was being taken to school in the back of my mother's Mini. The golden beams bursting from behind a plump Cumulus were mesmerising. It was the first time that I actually looked at a cloud and consciously wondered what it was. The crepuscular rays are what made me think about it. (My mother recently told me that when I looked up at the scene, I described it as 'silent thunder'.)

There are also such things as 'anti-crepuscular rays'. These appear to emanate not from the Sun, but from the opposite side of the sky – the 'antisolar point' – so cloudspotters have to have their backs to the Sun to see them. Like crepuscular rays, these are also the result of the contrast between light and shade in the air around clouds. Those regions where the sunlight is blocked, and so is not scattered by the general moisture in the air, stand out darker than the rest. The converging effect of anti-crepuscular rays is, again, due to perspective as the shafts of light recede into the distance.

Even when there is not enough moisture generally in the air to make the beams themselves visible, clouds can cast shadows on to other clouds. These sometimes appear as peculiar dark patches on a layer of cloud away from the Sun. The effect is particularly surprising when the Sun is low and the cloud casting the shadow is hidden over the horizon.

☁

RAINBOWS FALL WITHIN the group of optical effects explained by the way sunlight interacts with water droplets, such as raindrops of 1mm or so across, by reflecting back towards a cloudspotter facing away from the Sun.

Rainbows are most commonly seen in conjunction with the convection clouds like Cumulus congestus or Cumulonimbus. This is because those clouds are individual precipitating clouds, rather than expansive layers. With gaps in between, there is a fair chance of direct sunlight shining on to rainfall.

The sunlight passes into the raindrops and reflects off the inside of their far surfaces and back towards the Sun. The sunlight's constituent wavelengths are bent by different amounts as they pass into and out of the droplets, which has the effect of separating them. We see the differing wavelengths as distinct colours.

You can't get to the end of the rainbow, just as you can't stop the line of sparkling sunlight reflections on the sea surface from pointing towards you.

The rainbow that a cloudspotter sees standing in one position is never the same as that observed from another one. The droplets that are over in the direction of the arc – perhaps a half to one and a half miles away – each sparkle a bit of sunlight into his eyes. From the drops that fall through the sky off in some directions, it is the yellow-looking part of the spectrum that twinkles at the cloudspotter. From those in other directions, it is the violet, etc. This means that, should the observer change position, different raindrops will be the ones sparkling at him. Hopefully, this will help cloudspotters accept that it is a futile and, frankly, humiliating aspiration to seek the end of a rainbow. It is like driving a speedboat this way and that in an attempt to stop the line of glitters that the Sun casts on the sea's surface from pointing towards you.

Your rainbow is not my rainbow

Rainbows may be the most familiar of the sky's optical delights, but how many of us notice the finer points of their appearance? How many realise that the sky within the bow is brighter than that outside it? How many have spotted, on occasion, a secondary bow, outside the primary one, dimmer than it, and with the reverse order of colours? How many have seen 'Alexander's Dark Band'? This is not a goth group from Middlesbrough, but the name for the dark region between the primary and secondary rainbows. It was Alexander of Aphrodisias who first described it in around AD200. And how many of us have seen the faint blue/purple arcs that can

The gentle waves of Cirrostratus undulatus, ruined by the presence of CCTV.

appear just inside a bright primary rainbow? They are called 'supernumerary bows' – not a name to say in a hurry – and result from the 'interference' of light waves overlapping when they are slightly out of phase from each other, as they emerge from different parts of the raindrops. The principle of light-wave interference is discussed further below.

Cloudbows can appear, on rare occasions, when the sunlight falls on the much smaller water droplets in a cloud layer. As with rainbows, the Sun must be shining from behind the cloudspotter. The cloudbow arc has the same colours as the rainbow but they are softer and more diffuse, and the arc as a whole is a lot broader.

Fogbows appear when sunlight shines through a gap in the fog, from behind the cloudspotter. Rather than being coloured, these very broad arcs appear as ghostly, diffuse rings of white light. The particularly small droplets of the fog (perhaps just 0.02mm across) and their consistent size causes the light waves to interfere with each other as they bounce back towards the cloudspotter.

Closely related to the fogbow – though not fully understood – is a more colourful light effect called a 'glory'. It can appear around the shadows cast by cloudspotters when they are looking down on to cloud or fog, with the Sun behind them. It is seen more often by mountaineering cloudspotters, obviously, than by valley-dwelling ones.

Around the head of the cloudspotter's shadow, which is cast upon the cloud and usually distorted considerably by perspective, is a nimbus-like spectrum of colours. The eerie apparition is sometimes known as the 'Brocken spectre'. It is named after the Brocken, the highest peak in the Harz mountain range of Germany, which is shrouded in mists and fogs on up to three hundred days of the year, and where the glory can be seen quite often.

A cloudspotter who has companions with him when he observes this light effect will notice that the shadows of the others do not have glories around them. Each, in turn, will see the glory only around their own shadow. Clearly this is the most egotistical of the optical phenomena.

Those who can't be bothered to climb the Brocken – or any other mountains for that matter – need not live without glory. The same effect can sometimes be seen from the window of an aircraft. If there is the right sort of cloud layer below (very small droplets, of a very even size), the plane's shadow can be surrounded by a multi-coloured glory.

☁

THE FINAL GROUP OF optical effects results from sunlight observed through thin layers of water particles (be they droplets or ice crystals), which are very small (around one or two hundredths of a millimetre).

An example of this type of effect is the coronae that can appear around the Sun or Moon when seen through 'young' clouds – ones that have just formed and have an especially uniform size of water particles.

Coronae are most frequently observed through thin, newly formed layers of the mid-level Altostratus clouds, but can sometimes be seen in conjunction with Altocumulus, Cirrocumulus and Cirrus. More commonly noticed are the ones that appear around the Moon since this is, of course, not as dazzling as the Sun, and so can be looked at directly. A well-formed corona consists not of a ring (like the halos, mentioned earlier) but a central white disc, or 'aureole' – only a few times as wide as the Moon – with rings of colour around it. The order of colours tends to be a yellow-white disc (with the Moon at its centre) with a brown-red outer edge, followed by a faint spectrum of blue, green and then red bands. On occasions, there are further coloured bands appearing outside these.

Coronae are caused by the way the sunlight (either direct, or reflected off the Moon) bends as it passes around the tiny obstacles of a cloud's droplets or ice particles. Since the light does not need to pass through them, they tend to be much brighter than halo phenomena and rainbows (where the light is weakened, because it passes through the particles). This is why, even in the dim light of the Moon, they are often bright enough for cloudspotters to be able to see the colours around the aureole.

The diameter of a corona varies with the size of the cloud's droplets (the bigger the droplets, the smaller the corona), so it can sometimes be seen to grow and shrink as different layers of thin cloud are blown in front of a full Moon.

The light effects are the result of the droplets or crystals acting as tiny obstacles to the passage of light. This blocking of light only has a noticeable effect on the colours observed when the particles are very small. Just as the waves on the sea bend around an obstacle like the end of a pier, so do the light waves bend around the cloud particles. It is the fact that their constituent wavelengths of sunlight are bent to different degrees as they pass the particles, and the way the light waves from either side of each droplet interfere with each other, that cause the rings of colour around the corona's central bright disc.

The corona only looks sharp with pure colours when the layer of cloud is thin; whereas, through a thicker cloud, the light becomes diffuse as it passes around many particles on its route to the cloudspotter.

Since coronae can be explained by the sunlight bending around the obstacles, it should not surprise the keen-witted cloudspotter that they can form in droplet or ice-particle clouds and even clouds of opaque particles, such as pollen, blown up into the atmosphere by high winds. The ash and hydrated sulphate droplets that end up high in the atmosphere after volcanic eruptions can result in a corona known as a 'Bishop's ring', after Sereno Bishop, who first identified it in Honolulu after the eruption of Krakatoa in 1883.

The final example of clouds' optical effects – and one of the most beautiful – is closely related to coronae and is called 'iridescence' or 'irisation'. Iris was the Greek goddess of the rainbow, who was used by Zeus and Hera to pass messages and commandments down to mortals below. But her name has come to signify not rainbows, but the beautiful fringes of mother-of-pearl colours that can appear around the edges of middle and high clouds.

The bands of colours are effectively parts of coronae. They are usually in wavy bands because the droplets or ice particles become smaller towards the edges of clouds, as they evaporate away into the surrounding air.

Irisation often appears around wave clouds like the UFO-shaped Altocumulus lenticularis. The droplets form at one side of the cloud and evaporate away at the other, as they move in the air stream that causes these clouds. As a result they always remain very small, since they never have time to combine and grow into larger droplets. The fact that the Sun itself can be obscured by the cloud, making it less dazzling, means that irisation is more commonly observed than some of the other optical effects.

Whilst Iris may have given her name to a phenomenon quite different from the rainbow of which she was goddess, luckily for her the particularly intense, pure colours of irisation make this an even more beguiling and beautiful optical effect.

EVERY CLOUD HAS ITS MOMENT in the Sun. Each plays with the light for amusement. Some cast their shadows, like Chinese puppets, on to their neighbours, while others just break the rays with their forms, like fingers through a torch beam. Some separate and combine the spectrum as their droplets reflect light back to anyone who cares enough to notice, while others prefer to express themselves with the raindrops they cast, and no doubt they watch with satisfaction as the drops refract and reflect the spectrum into the arc of a rainbow. Each cloud has its own way of distorting the light, teasing out the waves and re-combining them. But, of all the optical effects, I can't help preferring the altogether subtler sparkles of the Cirrostratus cloud's halo phenomena.

Maybe this is because the subtle, milky layer of ice crystals seems the least noticed of all the clouds. I get the feeling this silent veil doesn't particularly mind that no one spots it. The Cirrostratus has no need to shout about its achievements. With its gentle cascade of ice prisms, this cloud is clearly content in the knowledge that it has displayed colours brighter than the rainbow's own, that it has mocked our one-and-only Sun with imitation sundogs and, of course, that it has changed the course of human history.

Oh yes, I almost forgot to say how.

Whilst the '*hoc signo victor eris*' writing in the sky was clearly by the hand of God, the sign that appeared with it (if the *labarum* symbols on some Roman coinage commemorating the victory are indicative) were, without a cloud-shadow of a doubt, by the gentle icy hand of the Cirrostratus cloud.

A rare coin – the 'Spes public' – struck in Constantinople in AD327, shows a particularly clear depiction of the military standard that became the established one for the Roman army, following the vision accorded to Constantine and his troops fifteen years before. It shows the *labarum* above a banner with three circles on it.

When the Sun happens to be at an elevation of 22° from the horizon, the smile of the circumzenithal arc can appear to touch the 46° halo. Were the cloud cover broken, so that just a part of the halo appeared below the arc, the effect is not a million miles away from the cross in Constantine's *labarum*. And the vertical line of the 'P', incorporated into the symbol? It is a sun pillar appearing below the Sun, of course. Three coins on the vellum below the symbol? Well, it goes without saying that they represent the Sun with sundogs, or mock suns, on either side of it.

Bronze Nummus, known as the 'Spes public', depicting the Roman army's military standard, inspired by Constantine's miraculous vision in the sky.

It is not inconceivable that, on the day before the Battle of Milvian Bridge, the ice crystals of a Cirrostratus in the skies above Constantine and his army were of the right sizes and orientations, and the Sun at the right elevation, for all four halo phenomena to appear at around the same time. Okay, not likely, but not inconceivable.

If they did, would they not look rather like different elements in the military standard depicted on the Roman coin?[4] Might Constantine have seen a cross in the form of a circumzenithal arc intersecting a broken 46° halo, with a vertical sun pillar down below? Might he have seen three balls of light – the Sun itself and mock suns to either side of it? Could that day, on the eve of the Battle of Milvian Bridge, have been one of the most important cloudspotting moments of all time?

Some, no doubt, will think it frankly ridiculous that a Cirrostratus cloud could be responsible for the spread of Christianity throughout the Western world. And, now that I think about it, they might be right: it could always have been a regular Cirrus cloud instead, for they too sometimes produce halo phenomena.

Not Forgetting...

ELEVEN

THE OTHER CLOUDS

The accessory clouds, supplementary features, and stratospheric and mesospheric clouds

T he ten main cloud types may get all the glory, but cloud-spotters should not forget the lesser-known characters in the cloud family.

Some are called 'accessory clouds'. They are the sidekicks, the wingman clouds – only appearing when one of the ten main types is around. Never far from the action, they sometimes become so swept up in the meteorological moment as to lose themselves into the body of the cloud they accompany.

Others are not considered to be clouds in their own right, merely 'supplementary features' of the big ten. However, cloud classification is a… well… nebulous pursuit, and decisions of what constitutes an actual cloud and what does not is often just a matter of convention amongst the world's meteorological community. It strikes me as a little unfair to deny some of these the puffy pride that must surely come with being acknowledged as an actual cloud, rather than merely a 'feature' on one.

Finally there are the 'stratospheric' and 'mesospheric' clouds, which are most definitely clouds in their own right. These form in the higher atmosphere, sometimes many, many miles above the troposphere, which is good enough for all the other clouds. From their impressive vantage-points, these mysterious characters look down on their brethren with lofty detachment.

ACCESSORY CLOUDS

PILEUS

The pileus is rather like a cloud haircut. It is a supercooled-droplet bouffant, worn exclusively by the fashionable Cumulus family. If cloudspotters keep their eyes to the sky, every once in a while they'll notice a Cumulonimbus or its younger brother, the Cumulus congestus, sporting this dashing hairstyle.

Pileus can form when the vertical convection currents of one of these towering clouds behave as an obstacle to a horizontal air stream above. The moist current of air is pushed upwards by the building cloud below. If conditions are right, the wave-like crest that forms as the air passes over the cloud is enough for some of its water vapour to condense into droplets as it cools.

This process of a pileus cloud's formation is very similar to the way in which wave clouds, such as the various lenticularis species, form orographically when air streams rise to pass over mountains. In the same way, the droplets in a pileus move through the cloud with the air, appearing at one side and disappearing at the other.

Justin and Shannon Moore (members 1477)

The pileus is like a cloud haircut.

This gives it a wonderful blow-dried appearance on the convection cloud's puffy pate.

But cloud fashion is more transient than most and, as the convection cloud continues to grow, its head soon pokes through the top of the pileus. Within a few short minutes, this cloud hair-do has slipped to its shoulder, leaving the poor convection cloud with all the indignity of a lopsided wig.

PANNUS

Pannus clouds are dark shreds of condensation, which form like ghostly apparitions in the saturated air of rainfall. When winds are gentle, they hang as dark patches below the belly of the rain cloud. In high winds, they have a more ragged appearance, scudding through the confusion of precipitation like spirits late for a haunting.

Cloudspotters will typically see pannus below Cumulonimbus, Nimbostratus, Cumulus congestus and thick Altostratus clouds. But they shouldn't cancel any important appointments to seek them out. These are not exactly the showstoppers of the cloud

Only ardent cloudspotters would ever notice pannus clouds.

world, looking about as exciting as shreds of Stratus cloud (which is exactly what they are). Most people never see a ghost. Even fewer ever notice a pannus.

But when cloudspotters do see pannus, they can be confident that, if the cloud above is not already raining, it will do so imminently. Unlike spreading, thickening Cirrus, which can be a presage of rain in a day or so, pannus are more of a three-minute warning. When the air becomes saturated with the first fall of rain, only the smallest updraught is required for some of it to condense into cloud droplets and form the pannus's ghostly shreds.

VELUM

Named after the Latin for a 'veil', the velum cloud is similar to the pileus, though it extends over a much larger area. It generally appears when a group of large Cumulus or Cumulonimbus clouds have the collective effect of lifting a stable layer of moist air. The velum doesn't necessarily form in a stream of air, like the pileus, but just hangs there, splayed out like a ballet dancer's tutu.

Often extending some way out from the convection clouds, this skirt can be visible even as the clouds rise through it towards the heavens. In fact, so stable is the velum layer that it often remains hanging in the air well after the convection clouds themselves have taken their bows and left the stage.

SUPPLEMENTARY FEATURES

TUBA

A tuba is a cloud extending its finger towards Earth. Young children often yearn to reach up and touch the soft mounds of a fair-weather Cumulus. Who can blame a cloud for trying to see what the ground feels like? It has to get worked up into a vigorous spin, however, before it can summon the energy to do so.

In and around the intense downdraughts associated with large Cumulonimbus and Cumulus congestus clouds, a vortex of swirling air can develop like that in water draining down a plug

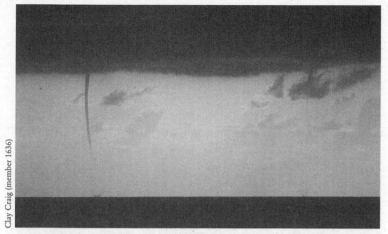

Clay Craig (member 1636)

Tuba – when clouds want to know what the surface feels like.

hole. The air tends to be flung out from the centre, as a result of all the spinning, which amounts to the same thing as a drop in the air pressure. When air decreases in pressure, it cools, and this can be enough for some of its water vapour to combine into droplets.

A tuba is thus a column or cone of cloud extending down the middle of one of these vortices. It is the first sign of a developing waterspout, landspout or – when the cloud is part of a particularly violent multicell or supercell storm – a tornado.

They do not always end up reaching the Earth's surface, however. More often than not, the cloud loses heart before touching terra firma. Maybe it knows that, in another guise – on a different day – it might reappear not as a convection cloud, but as fog or mist. Then, of course, it would end up hugging the ground until it is heartily sick of it.

INCUS

The ice-particle canopy that spreads sideways at the top of a Cumulonimbus is known as an incus. Not all thunderclouds expand laterally like this – only those that grow tall enough to encounter a strong temperature inversion like the tropopause (the region at the top of the troposphere that acts as a thermal lid on the convective rising of air).

Ashley Gibbs (member 563)

Incus – the anvil of ice crystals that spreads out above a Cumulonimbus.

When they do expand, however, Cumulonimbus develop into the shape of an anvil, which is what 'incus' means in Latin. Being so integral to the overall structure of the thundercloud, the incus doesn't really feel like a separate cloud – unless, that is, the Cumulonimbus rains itself out and the high plume of ice crystals is left behind.

The anvil is presumably what the Norse god Thor slams his thunder hammer on when he's having a bad day. As everyone knows, this is what causes peals of thunder.

MAMMA

The mamma cloud formations, sometimes known as 'mammatus', are named after the Latin for 'breasts'. These can appear on the underside of a number of different cloud types – Cirrus, Cirrocumulus, Altocumulus, Altostratus, Stratocumulus and Cumulonimbus – and at their most dramatic look like a field of smooth, globular udders.

They are at their most impressive when wed to a mighty Cumulonimbus. Forming on the underside of its incus, mamma appear when the top of the anvil cools, by radiating heat up into

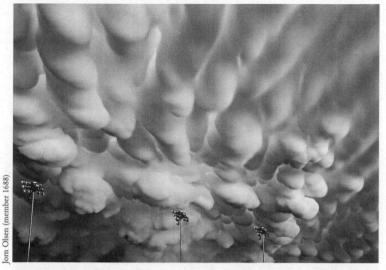

Jorn Olsen (member 1688)

Mamma – the udder-like protrusions that can hang below a Cumulonimbus anvil.

the atmosphere, and parts of it sink into the air below. When this air is relatively warm and humid, some of its water vapour condenses into cloud droplets as it mixes with the cold air. The process is like the reverse of convection currents forming into Cumulus clouds: rather than air warming at ground level and rising to form clouds, here air is cooling at the top of the troposphere and sinking to form them.

Mamma tend to be far less dramatic on the other cloud types. On the whole they are only plump, full and abundant when there is a mighty thunderstorm in the vicinity. The more powerful the Cumulonimbus, the more buxom the mamma.

ARCUS

Many of the accessory clouds and supplementary features appear in and around the rather egotistical Cumulonimbus, and the arcus is no exception. This is the low, horizontal leading edge of a storm that sweeps ahead of it and appears as a thick, ominous-looking shelf of cloud.

Arcus are generally associated with Cumulonimbus that have combined into a multicell, or even supercell, formation. The strong

Mike Hollingshead (member 1666)

A shelf of cloud, called an arcus, can appear at the base of a Cumulonimbus.

downdraughts of cool air are what cause it: these spread out as they reach the ground and push ahead of the storm, burrowing under the warmer air around and lifting it to form cloud.

The arcus is like a shelf, integrated into the confusion of weather at the base of the storm.

VIRGA AND PRAECIPITATIO

A cloud is described as praecipitatio when it has precipitation falling from it – be it rain, snow, sleet, hail, snow grains, ice pellets or cats and dogs – but only if this actually reaches the ground.

If, on the way down, the precipitation passes into a region of air that is warm and/or dry enough, it can evaporate away before it has a chance to touch down. This, as one would expect, is more often the case with the higher clouds, since there is so much air to fall through.

High clouds often produce ice crystals that fall some distance before evaporating. The falling, dwindling crystals are called virga and appear as wispy tendrils below the body of the cloud, sometimes giving it the appearance of a jellyfish. When virga

Above: Keith Epps (member 868)
Below: David Foster (member 1157)

ABOVE: Virga hanging below Altocumulus clouds make them look like a shoal of celestial jellyfish. **RIGHT:** A 'fallstreak hole' cut from a layer of Cirrocumulus.

appear below lower clouds, they are more commonly composed of evaporating raindrops. Either way, the wavy or slanting appearance results from the varying wind currents encountered through the precipitation's descent.

Cirrus clouds themselves are much like virga, for they are nothing more than fallstreaks of ice crystals. Whilst it would therefore be tautologous to talk of virga falling from Cirrus clouds, they do commonly appear beneath Cumulonimbus, Cumulus, Stratocumulus, Nimbostratus, Altocumulus, Altostratus and Cirrocumulus.

In the latter two cases, they can sometimes appear in a 'fallstreak hole'. This is a round gap that opens in a layer of supercooled liquid droplets when a region starts to freeze into ice crystals. As they grow, the crystals fall as virga, leaving a dramatic hole behind. This phenomenon is, in fact, more common than you might think.

☁

STRATOSPHERIC AND MESOSPHERIC CLOUDS

NACREOUS CLOUDS

If Cirrus are the most beautiful of the ten common clouds, then the rare nacreous clouds have to be the most beautiful of them all. Also known as 'mother-of-pearl' clouds, they stand apart from others both because they are the most colourful – displaying stunning pastel hues of pinks, blues and yellows – and because they form so high in the sky. At between ten and twenty miles up, they are not even in the same region of the atmosphere as the common clouds, which tend to form no higher than five or six miles in those parts of the world where nacreous clouds can be seen.

These are mostly at latitudes greater than 50° in both the Northern and Southern Hemispheres (they are more common in the south, though, for some reason, more colourful in the north). Composed of very small ice crystals (around 0.002mm across) and forming at temperatures of around −85°C, nacreous are visible in the twilight hours around sunrise and sunset. Whilst the rest of the sky is dark, and any of the lower cloud formations are in shadow, they are caught by the light of the Sun just over the horizon, and exhibit fantastic iridescent ripples of milky colours.

These appear in the same way that irisation, or iridescence, does when sunlight shines through the edges of the lower clouds. The nacreous cloud's ice crystals cause interference patterns in the sunlight passing around them. The effect depends upon the particles being very small and of a generally uniform size, as well as the cloud layer being thin. These factors are very much the case with this cloud, which is why its colours are so dramatic.

Nacreous are wave clouds – like extremely high versions of the lenticularis species of the main clouds that appear around mountain ranges. They form in the region of the atmosphere above the troposphere, which is known as the stratosphere. It is a pretty unusual place for clouds to find themselves. The temperature inversion of the tropopause usually halts the ascent of warm, moist air, so buoyant in the troposphere below.

The moisture of nacreous clouds manages to get past this lid because, like the lenticularis formations below, it is forced up by the waves that develop in the lee of mountains. Most of the time these waves only tend to produce tropospheric clouds, such as Altocumulus lenticularis or Cirrocumulus lenticularis, but when the atmosphere is particularly stable, the oscillations are transmitted all the way up through the troposphere. On occasions, they can be strong enough to break through the tropopause and carry moisture into the stratosphere above.

Sadly, the most beautiful clouds in the sky are also the most destructive to our environment. It is believed that nacreous clouds have the effect of speeding up the depletion of the ozone layer. It's not really their fault: the CFC gases that we have released into the atmosphere from aerosol cans and refrigerators are what react with, and break down, the crucial ozone. But the ice crystals of these high clouds serve as catalysts that encourage the chemical reaction.

Nacreous clouds seem to be becoming more common in northern latitudes. Why this is so, remains a mystery.

NOCTILUCENT CLOUDS

Nacreous clouds may be high, but noctilucent clouds are positively stratospheric! No wait, they are even higher than that: noctilucent clouds form at the top of the mesosphere, the region *above* the stratosphere, between 30 and 50 miles up. This is the coldest part of the Earth's atmosphere, near the fringes of space, where temperatures can be as low as −125°C. Forming right up there, with ice particles around 0.001mm in diameter, noctilucent clouds have a quality unique amongst all the clouds – they are lit by sunlight in the middle of the night.

They do not display the varied colours of nacreous, but tend to be mostly a milky-blue and are so thin as to be only noticeable against a dark night sky when the Sun has set but still lights the upper reaches of the atmosphere. These conditions are met for the longest durations at latitudes above 50°, within a month or so of midsummer.

Since they are so far away from us, the formation of noctilucent clouds is something of a mystery. Actually, it's a lot of a mystery –

Lee Montgomerie (member 280)

Noctilucent clouds form between 30 and 50 miles up, and shine at night.

we've no clear idea how or why clouds should form so high up. Theirs is a region of the atmosphere too high to be reached by weather balloons, which have a ceiling of 20–25 miles, and yet below the lowest orbit of the US Space Shuttle, at around 100 miles. Besides being extremely cold, this region of the atmosphere is also incredibly dry – millions of times drier than the air over the Sahara Desert, according to NASA.

The first recorded observation of the mysterious noctilucent clouds was after the eruption of Krakatoa in Indonesia in 1883. Huge amounts of volcanic ash were spewed into the lower atmosphere, resulting in spectacular sunsets that made sky watching a world-wide obsession. The ash spread around the entire circumference of the globe, its progress giving scientists great insights into the movement of jet-stream air currents. Noctilucent clouds spotted around this time were thought to be no more than ash finding its way into the upper reaches of the atmosphere. But after the ash eventually dispersed, the clouds still appeared. Some scientists have speculated that the volcanic ash somehow made it all the way up into the mesosphere and acted as the seeds, on to which the ice particles froze.

We can only speculate as to what particles serve as the icing nuclei for present-day noctilucent clouds now that Krakatoa is history. It is unclear whether particles from the lower troposphere do find their way up there, or whether dust from meteorites breaking up in the outer atmosphere serves the purpose. What is certain is that not only have noctilucent clouds continued to appear but, within the last hundred years, they have been observed more and more frequently over wider regions of the world.

This has led some scientists to suggest that their more frequent appearance is related to global warming. It is now common knowledge that increased concentrations of greenhouse gases in the lower to middle atmosphere (below the level of noctilucent clouds) have a warming effect on ground temperatures by trapping more of the Earth's radiation in. The less-publicised flipside to this is that they also have a corresponding cooling effect on the rest of the atmosphere above.

Global warming that is not the result of the recent increase in greenhouse gases would not be expected to coincide with a cooling of the outer atmosphere. The increased frequency of noctilucent cloud sightings might just be the result of more people noticing them, as they have become better known. But it could, in fact, be one of the most visible indications of the extent to which recent warming of the planet is due to our activities.

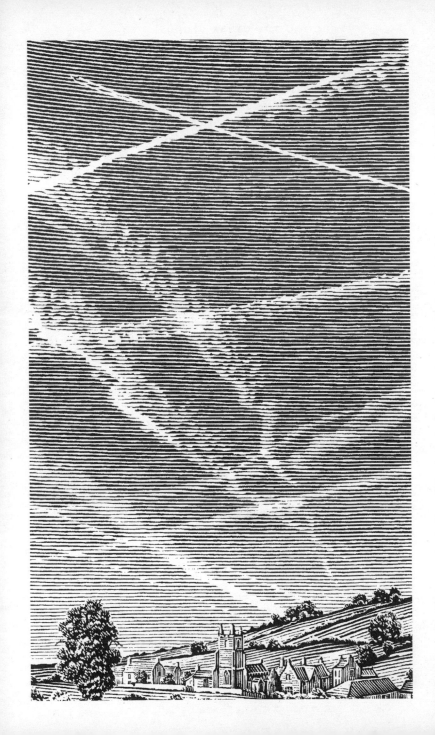

TWELVE

CONTRAILS

*The lines of condensation that
form behind high-altitude aircraft*

There must have been a cloudspotter or two amongst the cavemen.

I like to think that – some fifty thousand years ago – the odd enlightened Neanderthal stepped out of his cave on a beautiful morning, looked up at the sky, and grunted to his partner the equivalent of 'Honey, get out here quick – you're missing a fabulous display of Altocumulus stratiformis perlucidus!' It is, after all, tempting to speculate that cavemen would have looked up at exactly the same family of cloud types as we see above us today. Tempting, but wrong.

A new type of cloud has joined the family, you see. In fact, it is such a recent addition to the skies that it wasn't even around two hundred years ago when Luke Howard first named the clouds.

Arriving during the run-up to the First World War, the condensation trail, or 'contrail', is the straight, man-made cloud that forms in the wake of high-altitude aircraft. It is the new bastard son of the cloud family, though that term seems rather too derogatory. In this day and age, it is perhaps better just to say that it was not conceived in quite the same circumstances as its older stepbrothers and stepsisters.

It may sound surprising that contrails count as clouds in the first place. But the only difference between them and the others is that contrails are man-made – created from the water vapour in aeroplane exhausts, which is a by-product of engine combustion. Compared with the organic, chaotic forms of the natural clouds, these bright, sharp swathes of progress dissect our skies with a modernist linearity. They have the abstract simplicity of a Mondrian painting. I mean, of course, one from his mature years, such as *Vertical Composition with Blue and White*, and not the loose brush strokes of his youth, as exhibited in *The Red Cloud*. (Alas, I can't show the comparison, since I'm not allowed to reproduce them in black and white – you'll just have to look them up.)

The formation of contrails is akin to the way breath turns misty on a cold day. Whilst most clouds form when moist air cools as it rises, the hot gases in an aircraft's exhaust cool by mixing with the extremely cold air up at cruising altitude. This is usually between 28,000 and 40,000ft, where temperatures can be anywhere between –30°C and –60°C. The hot, moist gases in the aeroplane's exhaust cool very rapidly as they mix into this cold air, which can result in some of the moisture forming into droplets of water, which instantly freeze into ice crystals a wingspan or so behind the plane.

Planes do not always form contrails, however – even at cruising altitude. On one day, aircraft can be seen high in the air with no cloud forming behind them. On another, the cloud trail appears, only to disappear again a few hundred feet behind the plane. On some days, however, contrails hang around for hours, criss-crossing the blue with lines that are gradually distorted by high winds.

Their appearance and duration depend on the atmospheric conditions up at cruising altitude. When the surrounding air is warm enough (relatively speaking) and dry enough, the ice crystals can barely have formed before they evaporate into it.* When it is cold and moist enough, however, the air is contrail-friendly, and the ice crystals not only form easily but gather more moisture from the surrounding air, as they are spread by the wind, growing in size as they do so.

Different regions of air can vary greatly in humidity and temperature, so the trails sometimes appear as broken lines, as

---

* To be more precise, the crystals 'sublimate' into water vapour – they change into invisible gaseous water, without even melting into liquid droplets first.

When contrails persist, they can spread in the winds up at
the top of the troposphere.

planes pass from one pocket to another. In this way, contrails can
hint at moisture and temperature conditions at the top of the
troposphere and act as prognosticators of the weather.

In temperate latitudes, when contrails do not appear behind a
high-flying aircraft, or when they form only briefly, it is usually a
sign that the air at the top of the troposphere is sinking and/or it is
dry, which suggests that fine weather is likely to persist. When they
remain and spread out, it can indicate that the upper air is moist
and rising, which it does in advance of a warm front. In this way,
persistent contrails can warn cloudspotters – even before the
appearance of spreading Cirrus clouds – of an advancing warm
front, with precipitation in a day or so.

☁

AIRCRAFT EXHAUST contains a lot more than just water
vapour. The other ingredients include carbon dioxide, oxides of
sulphur and nitrogen, hydrocarbons, carbon monoxide, unburned
fuel and tiny particles of soot and metal. The particles have an
important role in the formation of the contrail, by acting as nuclei
on to which the water vapour can begin forming into droplets and
crystals.

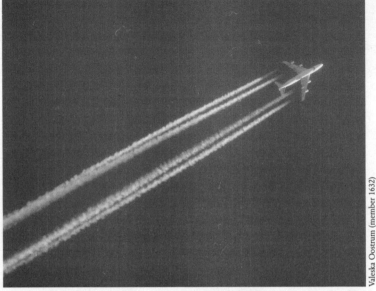

Valeska Oostrum (member 1632)

Contrails – the neatest and nastiest members of the cloud family.

It is not uncommon for the air at cruising altitude to be saturated enough to be on the point of forming droplets or ice crystals, except that there are not enough nuclei present on which it can get started. When this is the case, the introduction of a little more moisture and some particles can be enough to start a chain reaction. The exhaust from the aircraft is 'cloud seeding' by introducing the necessary ingredients for the air's water vapour to start combining into visible cloud particles.

On rare occasions, it is possible to see a dissipation trail, or 'distrail', which is like the inverse of a contrail. This is when the exhaust appears to cut a corridor of clear sky out of a pre-existing high layer of cloud, such as Cirrostratus or Cirrocumulus. Instead of a cloud in the aircraft's wake, there is a sharp gap in the cloud layer. This can happen when the aeroplane flies through, or just above, cloud, and is the result of one of three processes.

The heat of the exhaust can be enough to warm the cloud layer, so that some of its water particles evaporate away. The turbulence in the wake of the aircraft can mix drier surrounding air into the layer, having the same effect. Or the exhaust can seed the cloud layer – its particles encouraging the cloud's droplets to freeze and grow large enough to fall, so that they eventually evaporate away into warmer air below.

With contrails, we are unintentionally influencing the appearance of our cloudscapes. But the process of cloud seeding is not something that man just does inadvertently. For the last sixty years, scientists have experimented with introducing artificial nuclei into clouds to modify their behaviour. Cloud seeding was developed in an attempt to find a way to increase precipitation in drought-stricken regions, to reduce the severity of damaging hailstorms, to dissipate fog at airports and even to weaken the destructive nature of hurricanes.

☁

WHILST THESE MAY SEEM LAUDABLE reasons to mess with the clouds, seeding has also been employed for much more dubious ends.

The process was developed during the 1940s at the research laboratories of the General Electric Company in Schenectady, New York State. It was the brainchild of two scientists, Irving Langmuir and Vincent Schaefer. Langmuir, the director of the laboratories, was a well-respected chemist who had won the Nobel Prize in 1932; Schaefer was his research assistant, 25 years his junior.

During the Second World War, the laboratories were contracted to undertake military research for the US government, and Langmuir and Schaefer developed devices such as a smoke generator that could be used to screen military operations from the enemy. They also tried to solve the problem of icing on aircraft wings. This represented a significant danger to aviation, for the ice that can build up on a plane's wings as it flies through the sub-zero, supercooled regions of clouds can have a disastrous effect on its aerodynamics, by changing the wing shape and causing a potentially fatal loss of lift.

Their attentions soon shifted, however, from the challenge of *reducing* ice formation to that of *encouraging* it. Through their research on aeroplane wing icing, they learnt that crystal formation in clouds is one of the main processes by which their supercooled water droplets develop into large enough particles to fall as precipitation. It occurred to them that, were they able to encourage a cloud's droplets to freeze, they might make it more likely to rain.

Neither Langmuir nor Schaefer had any meteorological training. The first time they came across supercooled water droplets was in sub-zero temperatures on top of Mount Washington in New Hampshire. Both keen mountaineers, they made frequent visits to the weather observatory there during the war. Standing in the clouds at the summit, 1,919m above sea level, they were surprised to discover that the clouds consisted of droplets of liquid water, even at temperatures well below 0°C. The cloud droplets only formed into ice when they came into contact with solid objects, forming 'rime' – a little like an instant frost – on rocks, trees and buildings. The droplets were in a supercooled state – cold enough to form into ice crystals, but only when there was something for them to get started on.

Intrigued by the strange behaviour of supercooled cloud droplets, Langmuir and Schaefer installed an open-topped refrigerator in their laboratories to enable them to examine artificial clouds. By exhaling into the –20°C air of the refrigerated chamber, they could observe how the moisture of their breath would condense into clouds, whose droplets were in a supercooled state. The two scientists reasoned that if they could find nuclei that would encourage these floating droplets to form into ice crystals, they might be able to introduce the same nuclei into real clouds and encourage their water to fall as snow or rain.

☁

WHILST THE TWO RESEARCHERS had a clear idea of the theory of cloud seeding, the practicalities of finding particles that would act as icing nuclei were to prove a challenge. Langmuir and Schaefer tried introducing all sorts of additives into their cold chamber to encourage their supercooled breath clouds to freeze. These included soot, volcanic ash, sulphur, silicates and finely ground graphite. They were the sort of particles already present in the atmosphere, their reasoning being that some of these might act as the icing nuclei on to which cloud droplets freeze in the natural environment. However, none of the nuclei they introduced made any significant difference to the supercooled droplets of their breath clouds. Not only were they beginning to wonder if they would ever be able to encourage icing, they were running out of breath.

Then, one day when Langmuir was away from the laboratory, Schaefer made a breakthrough. He had decided to make the air inside the refrigerated chamber even colder by adding a block of dry ice (frozen CO_2, which has a temperature of around *Cloudbusters* $-78°C$). No sooner had he introduced it into the chamber, *are go!* than his breath cloud began to shimmer and sparkle in the light. It had instantly turned into crystals, which fell to the floor with the same dendrite shapes found in natural snow. By lowering the temperature to well below $-20°C$, Schaefer had made the supercooled droplets freeze even without icing nuclei.

The two scientists soon discovered that the critical temperature at which supercooled droplets freeze without any nuclei present is $-40°C$. It occurred to them that if they introduced pellets of dry ice into a cloud, they might be able to cool its droplets sufficiently to encourage them to form into snowflakes that would start to fall as precipitation. The search for icing nuclei was put on the back burner. Langmuir and Schaefer wanted to see what would happen if they dropped dry ice into a real cloud.

☁

ON 13 NOVEMBER 1946, Schaefer flew over a bank of supercooled Stratus cloud above Pittsfield, Massachusetts, and sprayed three pounds of finely ground dry ice from the aircraft. Within five minutes, snowflakes formed in the part of the cloud that had been sprayed, which fell about a thousand metres before evaporating into the warmer air below. A hole was left in the cloud where its supercooled droplets had frozen and fallen away. To prove that the hole was not something that might have happened naturally, a later test, before an audience of spectators, involved using dry ice to cut the General Electric logo from a bank of supercooled Stratus cloud.

These early test flights attracted a frenzy of publicity. The possibility of modifying the behaviour of clouds seemed tantalisingly close. Langmuir was outspoken in his enthusiasm. He eulogised to the press about the potential of their discovery: it would only be a matter of time before they could increase rainfall to banish drought. If they learnt to increase the freezing of droplets in storm clouds, they could alter their dynamics to reduce the size of hailstones, saving precious crops. He even proposed using it to tweak the cyclonic swirl of hurricanes so as to divert them away from populated areas.

To the meteorological establishment, Langmuir's promises seemed poorly founded. Conscious of the public's tendency to ridicule their efforts in the complicated task of weather prediction, they feared this outsider's claims would damage their already precarious reputation. Bitter arguments soon developed between Langmuir and meteorologists in the scientific press. Even if it were possible to affect the behaviour of clouds with dry ice, the meteorologists argued, it was by no means clear that it would be in any way economically viable. To seed clouds on a significant scale would be so expensive as to outweigh the benefits.

Langmuir and Schaefer were not to be discouraged, however, and their research took a step forward with the help of a third scientist, called Bernard Vonnegut, who had joined the General Electric labs.

Vonnegut was convinced that it would be possible to find a chemical to act as icing nuclei and encourage droplets to freeze when temperatures were above the critical $-40°C$. He identified chemicals with a similar crystalline structure to ice, believing that water would be likely to freeze on to something similar to it.

Consulting X-ray crystallography tables, he found that silver iodide was a possible contender. Indeed, when he blew a smoke of silver iodide crystals into the supercooled cloud in the refrigerated chamber, the results were dramatic. The freezing was instantaneous. The cloud formed into ice crystals and fell to the bottom of the

CLOCKWISE FROM TOP LEFT: Irving Langmuir, Bernard Vonnegut and Vincent Schaefer, the fathers of cloud seeding, in front of their refrigerated chamber in the research laboratories of General Electric.

chamber as snow. The silver iodide acted as nuclei on to which the supercooled droplets would start to freeze even at temperatures as high as –4°C. By now, those outside General Electric were beginning to take an interest in cloud seeding and, in 1947, the funding for the cloud seeding research at the company was taken over by the US government. Under the name Project Cirrus, research continued at General Electric until it was transferred in the 1950s to the Naval Weapons Center in China Lake, California.

Wait a minute – Naval *Weapons* Center?

☁

HISTORY HAS SHOWN AGAIN and again that weather can have a greater influence on the outcome of battles than military strength. Take the Persian Wars of the fifth century BC. When King Darius set out to strengthen the Persian Empire by invading Greece, his forces were overwhelmingly stronger than those of the Greeks – yet he was constantly hindered by the weather. On one occasion, a surprise attack on the Greeks was foiled when a tremendous storm destroyed practically the whole squadron of Persian ships.

On the eve of the Battle of Waterloo in 1815, it rained heavily on the French and Allied forces. Though the French outnumbered Wellington's troops, Napoleon was unable to launch his attack on the morning of the battle until the ground had dried enough for him to move his artillery into position. The delay allowed Prussian forces to reach the battlefield in time to support the Allies and ensure the French defeat.

Weather was also critical to the success of the D-Day landings of 1944. The Allies knew they would need a combination of settled, clear weather, suitable tides and a moonlit night if their invasion was to stand a chance. The tides and Moon suggested 5 June as the best night, but in the preceding days the weather was looking typically unsettled. Senior meteorologists at the Met Office, the Naval Meteorological Service and the Weather Service of the United States Army Air Force all worked to advise the Supreme Commander of the Allied forces on a suitable date. In what was one of the most important weather forecasts of all time, they found evidence that suggested a window of calm weather on 6 June. The date of the landing was changed by a day, and you know the rest...

It is not hard to see the huge military advantage a nation would obtain through being able to control the weather on the battlefield. So when the US intelligence community started reporting in the late 1950s that the Soviets were experimenting in weather modification too, the administration became concerned that if they didn't master it, they would be beaten to it. In 1957, a

Pilots cut the figure '4' out of a deck of supercooled cloud during Project Cirrus.

Presidential Advisory Committee on Weather Control concluded, 'Weather modification could become a more important weapon than the atom bomb.'[1] It seems that in parallel with the Cold War's much-publicised arms race, a secret weather race was also under way between the Americans and the Soviets. In the Sixties, two years after the outbreak of the Vietnam War, the Americans took the opportunity, under the veil of utmost secrecy, to try out cloud seeding in a combat environment.

ON 3 JULY 1972, the *New York Times* ran a front-page article by the Pulitzer Prize-winning journalist Seymour Hersh, exposing a covert operation conducted by the White House and the CIA during the Vietnam War.[2] Hersh claimed that the Americans had, for the past seven years, been spraying chemicals into the clouds over Laos, Vietnam and Cambodia in order to make them rain.

These were monsoon regions. During the wet season, the rain made the network of jungle paths, known as the Ho Chi Minh Trail, so muddy that they became virtually impassable. The US

Administration was well aware of the importance of the Trail for the Viet Cong and the North Vietnamese Army. Winding from North Vietnam, through Laos and Cambodia, into South Vietnam, these were vital supply routes for the enemy forces. They realised that if they could increase rainfall over the route at the start and the end of the monsoon season, they could extend its duration and disrupt enemy movements further.

Hersh revealed that the US Administration's first use of cloud seeding had been in South Vietnam – somewhat bizarrely, for the *Clouds and* purposes of crowd dispersal. Demonstrations were *the Vietnam* becoming increasingly problematic for the American-*War* backed Diem regime in the south. 'The... regime was having all that trouble with the Buddhists,' a CIA source told Hersh. 'They would just stand around during demonstrations when the police threw tear gas at them, but we noticed that when the rains came they wouldn't stay on. The agency got an Air America Beechcraft and had it rigged up with silver iodide. There was another demonstration and we seeded the area. It rained.'

The US then embarked on a top-secret programme of intensive cloud seeding tests over a section of the Annam Mountain Range, mostly in Laos, between 29 September and 27 October 1966. The project was given the codename 'Project Popeye'.[3] So sensitive was the project considered to be that, besides the military, only the President, the Defense Secretary, the Secretary of State, and the Director of the CIA were kept informed.[4]

In 56 seeding flights, 85 per cent of the clouds were considered to have reacted favourably, and the Commander in Chief Pacific reported to the Joint Chiefs of Staff that 'cloud seeding to induce additional rain over infiltration routes in Laos could be used as a valuable tactical weapon.'[5] Operational cloud seeding commenced on 20 May 1967 and continued for six years over parts of Laos, North Vietnam, South Vietnam and Cambodia, at an estimated annual cost of around $3.6 million a year. It is impossible to say whether it was really successful in increasing rainfall, since no systematic assessment of precipitation was made after the initial test phase, which itself could not be considered to be statistically rigorous. Nevertheless, the Defense Intelligence Agency later

The *New York Times*, 3 July 1972,
published Seymour Hersh's front-page exposé
of the use of cloud seeding during the Vietnam War.

estimated that the seeding activities increased rainfall by up to 30 per cent in limited areas.[6]

Whilst the story had originally broken in a column by Jack Anderson the year before, Hersh's front-page exposé brought it fully into the public eye. It caused an outcry, leading to awkward and embarrassing questions being asked in the US Senate about the military use of weather modification in Vietnam. These were initially met with evasive answers, but led to the Senate passing a resolution urging President Nixon to initiate negotiations towards a treaty against the manipulation of the environment for the purposes of war.

On 18 May 1977, under the presidency of Gerald Ford, a multilateral 'Convention on the Prohibition of Military or Any Other Hostile Use of Environmental Modification Techniques' (ENMOD) was tabled in Geneva. Signatories at the time included the US, the USSR and forty other nations. The convention, still in place today, was supposed to prevent countries from seeking to manipulate the weather for the purposes of warfare.

☁

BUT THE ENMOD AGREEMENT was vaguely worded. It only prohibits the military use of environmental modification techniques that could result in 'widespread, long-lasting, or severe effects'. By interpreting these constraints as it sees fit, the US government has continued to investigate the military potential of weather control. As recently as 1996, following the re-election of President Bill Clinton, a study was submitted to the Chief of Staff of the US Air Force entitled *Weather as a Force Multiplier: Owning the Weather in 2025*.[7] It is a chilling report, presenting proposals on how, by 2025, the US Air Force could exploit emerging technologies to 'own the weather' as a weapon of war.

Compiled by seven military officers in response to a directive by the Chief of Staff calling for an examination of 'the concepts, capabilities, and technologies the United States will need to remain the dominant air and space force in the future', the 44-page study was carefully worded to appear to be in accord with the ENMOD agreement. It claimed that its scope was limited to 'localised and short-term forms of weather modification'.

Owning the Weather in 2025 paints a terrifying picture of future warfare, in which the military are able to generate cloud cover and fog for the concealment of troop and equipment movement, to induce precipitation to flood enemy lines of communication, to deny precipitation to induce drought conditions, to steer storms over the enemy and even to induce lightning strikes on their targets. The report goes so far as to propose the use of nanotechnology to create controllable clouds of microscopic computer particles – buoyant in the air like real cloud particles and able to communicate with each other. 'The potential for psychological operations in many situations could be fantastic,' it enthuses.

This is taking warfare into frightening territory. As one, unnamed critic of the report's enthusiasts put it, 'They're like boys playing with a sharp stick, finding a sleeping bear and poking it in the butt to see what's going to happen.'[8] Clearly, the authors of the report couldn't care less about the environmental and human impact of using the weather as a weapon – they just want to make sure that someone else doesn't manage it first: 'While some segments of society will always be reluctant to examine controversial issues such as weather-modification,' it summarises, 'the tremendous military capabilities that could result from this field are ignored at our own peril.' It sounds like something from a science fiction novel. Indeed, Bernard Vonnegut, who had been so crucial in the early cloud seeding research at General Electric, was in fact the older brother of the science fiction author, Kurt Vonnegut.

Kurt worked briefly in the public relations department of General Electric, and was clearly inspired by his brother's research there. His bleak doomsday novel *Cat's Cradle* explored the repercussions of a chemical process that bore a striking resemblance to cloud seeding.

Dr Felix Hoenikker, a fictional Nobel laureate who helped develop the atomic bomb, creates a very unstable isotope of water, which he calls 'ice-nine'. Conceived as a way of stopping troops being bogged down on the battlefield, ice-nine has a freezing point of 50°C and only a tiny 'seed' of it thrown into the mud will start a chain reaction, causing all the moisture to freeze solid.

Hoenikker doesn't really think the matter through, though: this catalyst is so unstable that the reaction will continue until all the moisture on the planet freezes up. Just before his death, Hoenikker produces a tiny amount of ice-nine, which is subsequently *Sci-fi ice* divided up by his three children. The US and Soviet *horror* governments gain possession of it, as does the dictator of a tiny banana republic in the Caribbean. Needless to say, the whole thing ends in disaster when some of the ice-nine finds its way into the sea. This results in the oceans freezing over and the end of the world as we know it. Oops.

☁

OH DEAR, WE SEEM TO HAVE strayed a long way from Irving Langmuir's original idealistic aims for the uses of cloud seeding. Nevertheless, research into its peaceful applications has been actively pursued in many countries.

Interest reached its peak in the 1970s, when research funding in the USA alone was estimated at around $20 million per year. But the ambitious claims of enthusiasts as to its effectiveness have rarely been conclusively proven. Cloud seeding projects were often undertaken with little statistical methodology in place to evaluate their success. By the very nature of clouds, each one is unique, which makes it is impossible to find a 'control' – a cloud exactly the same as the seeded one – with which to compare its behaviour.

The inherent difficulties in proving that cloud seeding actually worked eventually resulted in the US research investment dwindling to $500,000 in the 1980s. It has continued to decline ever since.

One problem is that it is very hard to judge the correct amount of nuclei to introduce into a cloud to achieve the desired result. When attempting to increase precipitation, it is very easy to over-seed the cloud. Introducing too many nuclei results in so many ice particles or droplets forming and competing for the air's moisture that they all fail to grow large enough to fall as precipitation. This can actually reduce the cloud's tendency to precipitate. Indeed, many in the scientific community still doubt whether there is sufficient evidence to claim cloud seeding can effectively enhance precipitation.

Nevertheless, active programmes continue. In 1999, the World Meteorological Organisation reported that more than a hundred weather modification schemes were registered with them in 24 countries around the world. China is currently the most active, with an annual investment in weather modification estimated at more than $40 million.

The schemes fall into three main categories: fog dispersal, rain and snow enhancement and hail suppression. Of the three, fog *A better class* dispersal is generally considered the most effective. It is *of seeding* typically used at fog-prone airports and motorways. One programme has been employed successfully for several decades at Salt Lake City International Airport.

The dispersal of fog forming in temperatures above 0°C (known as warm fog) often involves using jet engines to warm the air. But when temperatures are below 0°C (known as cold fog), the fog droplets are encouraged to freeze into ice crystals, which fall to the ground. This can be done in two ways. Some projects involve seeding with silver iodide using rockets or airborne dispensers, whilst others, like that in operation at Salt Lake City, employ the introduction of dry ice to make the fog freeze without introducing icing nuclei.

The majority of seeding projects aimed at increasing rain or snowfall currently take place in the tropical semi-arid belts on either side of the Equator. Whilst many focus on the supercooled regions of clouds, aiming to increase droplet freezing, some seek to encourage precipitation from 'warm' clouds that do not contain significant supercooled regions. In these cases, salt particles, droplets of saline solution or even those of just water are introduced to encourage droplet growth by coalescence, or collision. Orographic clouds, which form around mountainous ranges, seem to respond best to seeding attempts, particularly when they contain a mixture of supercooled droplets and ice particles. They are also sensible clouds to focus on, since rain and snowfall on mountains can be stored in reservoirs and snow packs until needed.

Whilst it is generally agreed that the most cloud seeding can achieve is to increase the yield from clouds likely to rain anyway, the WMO claims, 'Statistical analyses of surface precipitation records from some long-term projects indicate that seasonal increases have been realised.' I think they're saying that it works.

Within the last ten years, crop damage caused by hail in the USA was estimated at around $2.3 billion annually.[9] Seeding programmes aimed at hail suppression are usually performed at the periphery of storm systems. Some are designed to encourage more hail embryos to form than would naturally do so, so that they compete with each other for the clouds' moisture and never get the opportunity to grow to a dangerous size. In other cases, the seeding programme aims to lower the height of the cloud system and reduce the trajectory in which the hailstones grow. Some simply try to make the clouds precipitate their moisture out before growing large enough to generate damaging hail.

The commercial firms offering these programmes say they can achieve large reductions in hail severity, but the scientific evidence to support them is generally considered inconclusive. In what seems to be a familiar hedging in the field of cloud seeding, it neither conclusively affirms nor denies the effectiveness of hail-suppression schemes.

☁

THE MAN WHO HAS BEEN the most flamboyant enthusiast of cloud seeding in recent decades is Yuri Luzhkov, the outspoken Mayor of Moscow. Ever since he was first elected to run the city in 1992, Luzhkov has shown a particular fondness for cloud seeding. But his interest has not been to increase rainfall – quite the opposite. Luzhkov uses cloud seeding to stop it from raining on his parades. A chemist in his younger days, the mayor is rather obsessed with the weather. He once became so infuriated with incorrect forecasts that he threatened to cancel the city's contract with the government meteorological service and start his own.

Luzhkov first employed cloud seeding in 1995 at Moscow's celebrations of the fiftieth anniversary of victory in the Second World War. Silver iodide was used in an attempt to encourage the clouds to rain before they reached the city. The parades were held in bright sunshine. In 1997, Luzhkov spent £550,000 in an attempt to guarantee rain-free days during Moscow's 850th anniversary celebrations. Eight planes were used throughout the three-day event to seed the clouds about sixty miles outside the city. No rain fell for the first two days, but as the outdoor closing ceremony approached on the third day, the weather turned.

By the time the event was in full swing, a steady downpour drenched the crowds. Acrobats and dancers performing on the soaking stage were slipping all over the place. Undeterred, the mayor continues his cloud seeding programmes before major events, and remains convinced of their success.

By contrast to all of this, aviation contrails are a way in which man continues to modify the clouds on a daily basis. Whilst the more sinister applications of cloud seeding make for worrying reading, there is growing scientific evidence that it is the impact of this inadvertent, continuous cloud seeding that we should really be concerned about.

Contrails sometimes take on a jagged appearance – like half of a zip – which is due to unstable air up at cloud level and the turbulence of the aircraft's wake.

☁

CONTRAILS PRESENT A BIT of a dilemma for cloudspotters. On the one hand, they are both interesting to observe and often very beautiful. When the air at cruising altitude is unstable, for instance, trails can form teeth of cloud below them, making them look like half of a zip – as if the aircraft is delicately undressing the heavens with its progress across the sky. At times, the trails can briefly take on a twisted appearance, like fusilli pasta. How this happens has yet to be explained. And, when conditions are right for them to spread out in the wind, contrails can develop into glorious interweaving lattices – the sharp lines of young trails dissecting the broad, diffuse swathes of mature ones.

On the other hand, there is growing evidence that the preponderance of contrails is having a significant impact on temperatures down on the ground, and yes – you guessed it – the overall effect seems to be a warming one.

Assessments of the environmental impact of aviation have

Mike Davies (member 1633)

When conditions are right, contrails can persist and spread out to lead to Cirrostratus layers that cover thousands of square miles.

traditionally concentrated on the contribution to global warming from the CO_2 in aircraft exhausts. Like all greenhouse gases present in the atmosphere, CO_2 has the effect of absorbing and radiating back some of the Earth's heat and slowing its cooling. Whilst aircraft emissions are estimated to account for two per cent *A heated* of all the CO_2 that man pumps into the atmosphere,[10] it *cloudspotting* seems likely that their environmental impact is greater than *debate* that of ground-based emissions, since they are released directly into the upper atmosphere. What might make contrail-loving cloudspotters think twice is that there is a growing consensus amongst scientists that the greatest environmental impact of aviation comes not from its greenhouse gas emissions but, in fact, from the clouds that flights create.

Clouds in general have a huge, if somewhat contradictory, effect on ground temperatures. It is well understood that when water is in the invisible, gaseous state of water vapour, it behaves as a greenhouse gas and tends to keep the Earth warm by trapping in its heat. In fact, water vapour is by far the most abundant greenhouse gas in the atmosphere, accounting for anywhere between 36 and 70 per cent of the greenhouse effect. Much less straightforward is the effect water in the atmosphere has on global temperatures when it condenses into clouds of droplets or ice crystals.

On the one hand, clouds block some of the Sun's radiation by reflecting it away from the Earth – something that is familiar to sunbathers who have felt chilly as a cloud drifted in front of the Sun – and so they lead to a localised cooling at the Earth's surface. On the other hand, like water vapour and the other greenhouse gases, clouds also absorb some of the Earth's heat and re-radiate a portion of it back down – which is the reason why cloudy nights generally feel warmer than clear ones. In this way, clouds have the opposite effect of slowing the Earth's cooling.

Most types of cloud, which can be seen to block out some of the sunlight, have an overall cooling effect. This is not so, however, for many of the high, ice-particle clouds – the Cirrus, Cirrostratus and Cirrocumulus, known collectively as 'cirriform' clouds. When they are thin enough to let much of the sunlight through, the effect of their trapping the Earth's warmth in outweighs that of cooling.

And it is this behaviour of the cirriform clouds that has a bearing on the environmental effects of contrails.

As any cloudspotter will attest, they don't just hang around as sharp lines of cloud. In the contrail-friendly conditions of low temperatures and high humidity up at cruising altitude, a trail's ice crystals can spread out in the wind. By acting as condensation and icing nuclei, the tiny particles in the aircraft exhausts encourage the water vapour naturally present in the atmosphere to form into droplets or crystals. The contrails' ice crystals themselves also serve as nuclei, growing as they pick up water molecules. Over a matter of hours, contrails can spread to become several miles wide. Indeed, individual contrails have been observed to spread out until they cover areas as large as 8,000 square miles.[11] By seeding the saturated atmosphere, they act as catalysts that encourage the formation of thin cirriform clouds – the very clouds that are known to raise temperatures at ground level.

☁

THE TERRORIST ATTACKS on New York's World Trade Center on 11 September 2001 shed an unexpected light on the way contrails affect temperatures at the ground. For three days

Contrails and 9/11 following the tragedy, all commercial flights over the United States were cancelled. For the first time since the First World War, there was a brief but sustained period without contrails over the USA. In a report published in *Nature* in 2002,[12] meteorologists compared the ground temperatures across the 48 contiguous US states during those three contrail-free days with the equivalent ones in temperature records of the previous thirty years. They did indeed find a significant difference. Without the presence of contrails, the difference between the daytime and night-time temperatures, averaged across the USA, was found to be 1.1°C greater than normal. It appeared that the contrails – and any cirriform clouds that they gave rise to – reduced ground temperatures during the day and raised them at night.

Whilst 1.1°C is actually a dramatic impact on ground temperatures, the study didn't show conclusively that the contrails were leading to an overall rise in ground temperatures, thereby suggesting a contribution to global warming. More recent studies, however, have suggested exactly this.

One paper, published in 2004,[13] looked at the increase in observed cirriform clouds over the USA between the years 1974 and 1994. Since the average humidity at Cirrus level was shown to be constant across the USA during this period, it was concluded that the increase in air traffic and its resulting contrails had led to increasing cirriform cloud cover. Estimations of the expected warming effects of this increase were equivalent to 0.2–0.3°C per decade. Amazingly, the effect of the increase in cirriform clouds alone was considered sufficient to account for almost the entire rise in temperatures across the USA during the last 25 years. This is a major claim, for though it relates to localised warming effects, not global ones, the report suggests that the high clouds that develop from contrails are a huge contributor to surface warming.

Another key paper, published in 2003,[14] was equally sobering. Here, the scientists correlated the changing distribution of cirriform clouds over Europe from weather satellite images with precise records of the varying concentrations of air traffic during the same periods. The report concluded that the warming attributable to cirriform clouds appearing to develop as a result of air traffic was ten times greater than that expected to result from aviation CO_2 emissions.

Now, it is hard to make a meaningful comparison between the environmental impacts of such differing factors as, on the one hand, aircraft CO_2 emissions, which remain in the atmosphere for over a hundred years and have a cumulative and global effect on surface warming and, on the other hand, aviation-induced cloud cover, whose warming effects are both localised and temporary. But these studies suggest that aviation's contrails are leading to other high clouds that are a more significant factor in global warming than its CO_2 emissions.

Air traffic is estimated to be increasing by five per cent a year,[15] with most of the increase being in contrail-forming long-haul flights. Ironically, modern aircraft engines – designed to burn more efficiently and so emit less CO_2 – actually create more contrails.

☁

A TEAM OF SCIENTISTS at Imperial College in London has been looking at one possible way to reduce contrails: stopping aircraft from flying so high.

Using computer simulations designed for air-traffic management, they have considered the implications of imposing restrictions on European cruising altitudes to keep aircraft below contrail-forming levels.[16] One problem with such a system is that the lower an aeroplane flies, the denser the air it has to travel through and so the more fuel it needs to burn – something that has financial implications as well as those of increased greenhouse gas emissions.

So the team evaluated a system that imposed the highest possible 'contrail-free' ceiling on cruising altitudes, which could be calculated dynamically in response to changes in atmospheric temperature and humidity.

'If you had that cap on the flights in Europe –' explained Dr Bob Noland, one of the scientists behind the project, 'which would result in a four per cent increase in CO_2 emissions from increased fuel consumption – our conclusion was that the reduction in contrails would make it a good policy.' Their findings suggested that, though there would certainly be implementation difficulties, such as increased congestion and longer flight times, the system could reduce contrail formation by between 65 and 95 per cent, compared with just a four per cent rise in CO_2 emissions.

Without the contrails it seems that there would be a considerable reduction in the overall amount of thin, ground-warming cirriform clouds. 'The CO_2 emissions from aircraft,' says Noland, 'while significant and growing, are not going to make that much difference even if we cut them down, but if we reduce contrails by 90 per cent tomorrow – which we think is entirely feasible – you would get a major impact right away. Stopping the contrails would bring an immediate benefit.'

☁

Should we consider banning contrails from our skies?

CLOUDS ARE THE WILD CARDS in climate change predictions. No one knows what effect a warming globe will have on the extent and nature of our clouds, nor what feedback effects the changes in cloud cover will have on the world's 'radiation budget' – the degree to which the Earth tends to retain heat from the Sun. The past century has seen a rise in global temperatures of around 0.6°C and the majority of this has taken place within the last fifty years. That this is in part due to human factors is now accepted by the vast majority of the scientific community.

If we continue a 'business-as-usual' policy with regard to emitting greenhouse gases, scientists estimate that, within a hundred years, we will double the amount of CO_2 in the atmosphere compared with the pre-industrial level. The direct effect of this is estimated to be a 1°C hike in surface temperatures.

Ken Bushe (member 1525)

A contrail casts its shadow on to the Cirrostratus layer below.

While this in itself would not be a catastrophe, the potential knock-on effects might be. The 'feedback' effects might amplify the temperature rises.

There are generally considered to be three main feedback factors. The first is the amount of ice covering the globe: since ice reflects sunlight away more than land does, any reduction in ice coverage will tend to have a positive effect on global warming. The second is the amount of water vapour present in the atmosphere: this is water in an invisible gaseous state and, like CO_2, is a greenhouse gas that traps heat in like a blanket. If rising surface temperatures mean more evaporation of surface water, then this would also have a positive effect on global warming. And then there is the third feedback factor – the clouds.

Clouds are the trickiest of all because we are so ill equipped to anticipate what a rise in global temperatures would mean to the nature and extent of cloud cover. If it caused an increase in thicker clouds, this might be expected to counter global warming by reflecting more sunlight away. If it resulted just in larger areas of high, thin cloud cover, then it could be expected to have a warming effect that would exacerbate the situation.

Perhaps the most problematic effect, however, would be if a rise in global temperatures resulted in fewer clouds. Not only would this put The Cloud Appreciation Society in jeopardy, it could have a drastic effect on the range of surface temperatures.

If higher temperatures were to result in more clouds being 'burned off' or more of them 'raining out' and dissipating more quickly, they might be expected to lead to clearer skies. In fact, Bruce Wielicki of NASA's Langley Research Center has found that there is currently less cloud cover in the tropics than there was 25 years ago. He says that the rising air over the hot equatorial zones seems to have increased in strength, which might explain storm clouds raining out more quickly, leaving the rest of the tropics less cloudy. Whether this is a trend resulting from rising temperatures and whether it can be extrapolated to have a bearing on non-tropical regions is, as yet, unclear. More certain is that, with fewer clouds around, not only would their contribution to the amount of sunlight reflected away from the Earth reduce, so would their contribution to the blanket effect of trapping the Earth's heat in. Crucial to deciding the effect of a reduction in cloud cover is how the two contributions to the Earth's radiation budget compare.

It is time for some very broad guesstimates: cloud cover as a whole can be considered to reduce the amount of solar radiation absorbed by the Earth by an average of around $50W/m^2$, while the extent to which it stops heat escaping from the Earth is more like $30W/m^2$. If those estimates are correct, the net contribution of the planet's cloud cover to its radiation budget is an energy loss of $20W/m^2$. In other words, the planet is cooler as a result of the clouds. Were they to disappear (all other factors remaining the same) the Earth would warm up even more.

Cloudspotters can tell that to the next person who complains about our fluffy friends.

☁

AS SCIENTISTS AND POLITICIANS continue to argue over the extent to which we are causing global warming and the curbs that are necessary on our emissions, I can't help drawing a parallel with Blaise Pascal, the seventeenth-century French philosopher, and his argument for the existence of God.

We cannot be sure whether God exists or not, Pascal argued, so we should consider our belief in his existence as a bet – one with unbelievably high stakes. If we believe he exists and, on our death, it turns out that he does, hooray! We go to Heaven. If we deny his existence, only to discover that he does exist after all, we face eternal damnation. Of course, if there is in fact no God, then it makes no difference either way, so surely, Pascal argued, any right-thinking person would lead a religious life.

The same holds true for global warming. Though the stakes are limited to mortal affairs, rather than the afterlife, they are equally momentous. The big unknown is our responsibility for the planet heating up. But the severe implications of it doing so – presumably amplified by feedback factors such as changes in cloud cover – mean the only sensible bet is to assume we are to blame and drastically change our behaviour.

☁

THE POET RUPERT BROOKE speculated before the First World War that 'the Dead die not', but ride the calm mid-heaven as clouds, and 'watch the moon, and the still-raging seas, and men, coming and going on the earth.'[17] This may indeed be so but, somehow, I can't imagine any of us end up as contrails.

It is hard to love these bastard sons of the cloud family. Some cloudspotters may, like me, feel ambivalent towards them, but how many can welcome these newcomers with open arms? They may cut dashing figures across the rosaceous autumn evening, but these icy swathes of progress must be the writing on the sky for cloudspotters and everyone else besides.

THIRTEEN

THE MORNING GLORY

The cloud that glider pilots surf

A few years ago, I was passing the time looking at the pictures in a cloud book[1] – which, I guess, is the cloudspotter's equivalent of reading *Heat* magazine – when I came across the photograph of a cloud that was unlike anything I'd seen before. The aerial shot showed an extremely long, smooth tube of low cloud that looked like a white roll of meringue and stretched from horizon to horizon, with clear skies ahead of it and behind. It had formed above an exotic-looking terrain of twisting rivers and mangrove swamps. I knew that it would be classified as a 'roll cloud' – a particular formation of the Stratocumulus genus – but it looked almost too sublime to be grouped with any of the common clouds. Indeed, the photograph's caption explained that it had a name all to itself – the 'Morning Glory', which 'conveys the feeling of elation which its passage arouses'.

A cloudspotter's life should not be spent looking at books. So I vowed, there and then, that I would find out where I'd be most likely to spot the Morning Glory, and go and see this beautiful cloud for myself.

☁

AND THEN I READ that it only forms in one of the most remote parts of Australia – the Gulf Savannah region of northern Queensland, which is pretty much as far from where I live as it is possible to go. Crossing the entire world in search of a cloud was surely too ridiculous a mission even for the most ardent cloudspotter. I realised that I might have been a little hasty with my vows.

The more I found out about this cloud, however, the more intrigued I became. I learnt that the Morning Glory can stretch 600 miles – as long as Britain – and moves at speeds of up to 35mph. Moreover, a small group of intrepid glider pilots travel thousands of miles across Australia, each year, in the hope of encountering it. They wait around, during the springtime months of September and October, in the tiny settlement of Burketown, where the cloud usually forms, with one mission: to 'soar' the Morning Glory. It is considered to be one of the most amazing gliding experiences, and one that can only be described as cloud surfing.

Suddenly, Australia didn't seem like such a long way to go after all: sitting around drinking beer with a bunch of Aussie glider pilots, waiting for the ultimate cloud to sweep overhead... it sounded like a fantastic Antipodean, aerial version of *The Big Wednesday* – surely, a very good reason for going to the other side of the world.

☁

AS I PREPARED FOR MY TRIP, I remembered the wealthy Victorian cloud enthusiast, the Hon. Ralph Abercromby, who had brought together the Cloud Committee of meteorologists that was to create the first *International Cloud Atlas* in 1896. Abercromby had spent several years in the late 1880s travelling the world in steamboats, trains and coaches in search of clouds.

He wrote a book of his travels, *Seas and Skies in Many Latitudes, or Wanderings in Search of Weather,*[2] which reads like a meteorological version of *Around the World in Eighty Days*. Less Phileas Fogg than Phileas 'Fog', Abercromby was curious to find out whether clouds differed around the world, and concluded that, by and large, they don't. A pioneer of meteorological photography, he recorded the formations he came across in remote regions. Many of his photographs were subsequently used to illustrate the *Cloud Atlas*, which he wrote with the Swedish meteorologist, Professor H. Hildebrand Hildebrandsson, in 1890, and which was a precursor to the international version.

In the weeks before my departure to Australia, I resisted the idea of growing a waxed moustache, like the impressive specimen sported by the intrepid Abercromby, deciding that it would make me the subject of ridicule amongst the Aussies. Nevertheless, I set off with words from his book ringing in my ears:

> *One of the Author's strongest wishes was to encounter a tropical hurricane, either on sea or land… But, though he selected the hurricane season for visiting Mauritius, and sailed all through the China seas in hope of meeting with a typhoon, he was not successful in his search.*[3]

Might not the same thing happen in my search for the Morning Glory? Clouds, after all, are the most chaotic of Nature's displays – the occurrence of even the common ones is hard to predict with any certainty. I'd heard of pilots driving the length of Australia to surf this spectacular cloud, only to head back a few weeks later, their glider having barely left the tarmac.

As the flight from London climbed through the deck of Stratocumulus that shrouded the city below, I feared that I was setting myself up for the most monumental cloud-anti-climax of all time. With this in mind, I did what only a cloudspotter would, when departing for an exotic location: I closed my eyes and prayed for clouds.

☁

BURKETOWN lies on the Albert River, about twenty miles inland from the enormous Gulf of Carpentaria, in the middle of Australia's northern coastline. With a population of just 178, it is not the sort of place you'd expect people to cross the world to visit – no more than a tiny cluster of lights in an immense expanse of

Russell White (member 23)

Burketown, Northern Queensland – just a little bit beyond
the middle of nowhere.

night from the window of the light aircraft that brought me from
Mount Isa, two hundred miles to the south. 'Yeah, it's real outback
here,' said Paul Poole, who runs the town's light-aircraft charter
firm, as we drove from the airstrip. 'It's one of the last untamed
areas of Australia.'

Poole operates flights between the various remote townships of
the Gulf of Carpentaria. With the immense distances involved and
the inhospitable savannah terrain, aircraft are the only sensible way
of getting around. The town relies on Poole's aircraft more than
ever during the December to February wet season, when this whole
flat region becomes flooded, rendering the dirt roads in and out of
it impassable. 'Once they complete the road from Mount Isa,'
Poole said, 'you won't recognise this place. The change will be

unbelievable and the Morning Glory will become the biggest thing in Australian gliding.'

With his girlfriend, Amanda, Poole also owns one of the few places to stay in Burketown. I was exhausted and crawling off to bed when they warned me that the cloud had passed over that very morning and so there was a strong possibility there would be another around dawn. 'It tends to work in a cycle,' Poole added. 'When the Glory comes, it's usually for a few mornings in a row, at around daybreak. The glider pilots get up at 4.30 or 5am to be ready.' Even though I'd been travelling for 42 hours, a lie-in was clearly out of the question.

I fell out of bed at five, when it was still dark, and drove their jeep to the salt flats north of the town. There I stood, surrounded by a swarm of bush flies and an endless expanse of flood plains, staring at the brightening sky.

The cloud normally approaches Burketown from the northeast, first appearing as a dark line on the distant horizon. But the orange, lilac and indigo hues of the sunrise stretched out across the flat savannah in a completely cloudless sky.

☁

ESTABLISHED IN THE 1860s as a supply town for cattle farmers, Burketown is situated on the natural dividing line between the northern wetlands and the southern grasslands of the Gulf region. It is the oldest settlement in the shire and has survived cyclones, epidemics and yellow fever. It looked exactly as an outback town should – metal houses, raised on rickety stilts, along red dusty roads.

Large grey cranes, known as Brolgas, ambled down the main road in the rising heat of the morning, and wallabies scampered into the bush as I drove by. The original 1865 customs house still stands, preserved as the town's only pub. There, I met old Frankie Wylie, who is such a regular fixture that he has his meals-on-wheels delivered to him at the bar.

Wylie seemed used to cheering up despondent cloudspotters. 'With this cloud you can't say when the time'll be,' he said.

Alas, no cloud on the horizon.

'Usually they say it'll come end of September, but then maybe it won't do till sometime in January. They don't know for sure – no one does.' He was sitting at the bar in a chair marked 'No parking' and held his 'stubbie' of beer in a leather-clad cooler that was inscribed with his name.

'First saw the Morning Glory when I moved here in '79,' he reminisced. ''Course you see clouds coming in everywhere around the world, but these fellas go upside down.' He made a rolling movement with his hands to *Consolation* show how the tube of cloud rotates away from its direction *in the pub* of travel. 'When you see them coming towards you, you think, hang on a minute mate, there's something wrong here –' (he shook his head, as if to sober himself) 'either my eyes are fading or I've had too much. It's something I'll never fully believe.'

On the pub's corkboard, beside photographs of locals holding up prize catches of Barramundi fish, were shots of the enormous cloud passing over the town. 'As it comes, it stirs all the dust and leaves and God knows what, but when it's over you the air will stop dead – no wind, no nothing. It's a weird experience. Why would it stop the wind like that?'

☁

DR DOUG CHRISTIE, of the Research School of Earth Sciences at the Australian National University in Canberra, knows the answer to questions like Wylie's, so the next day I went to the public telephone box to give him a call. The phone stands outside Burketown's post office, which has its own Morning Glory photos displayed above the counter. It struck me that for a town so far off the beaten track, the cloud is the equivalent of a visiting celebrity. Like a New York pizzeria proudly displaying signed photos of

Robert De Niro ordering a margherita, every public place in Burketown has its record of this famous spring visitor. With each photograph I came across, my desire to see the cloud for myself became more acute.

Christie is an expert in 'large amplitude atmospheric wave disturbances' – I suppose someone has to be – and is considered a world authority on the Morning Glory. In the 1970s, he *A call to* became interested in readings he'd picked up on the ultra- *Mr Morning* sensitive micro-barometer array at the university's research *Glory* station in central Australia. These he determined to be caused by very large individual waves of air, which he eventually traced to the Gulf of Carpentaria – four hundred miles to the north.

Since he first visited Burketown in 1980, Christie has conducted numerous experiments in this region and come up with the most accepted explanation for the phenomenon of the Morning Glory. He explained that the cloud forms in the middle of an enormous 'solitary wave' of air, which often seems to originate over the Cape York Peninsula – across the Gulf, to the northeast. This wave can travel as an independent crest, like an aerial version of the tidal bore

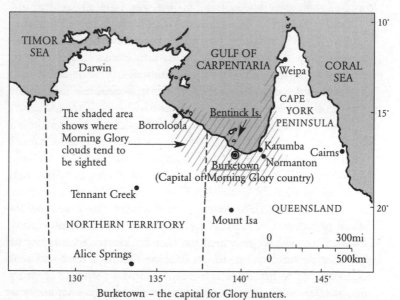

Burketown – the capital for Glory hunters.

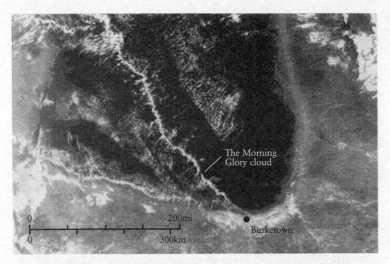

The Morning
Glory cloud

0 200mi
0 300km

Burketown

This satellite photograph, taken on 8 October 1992 (08:00 Local Time),
shows the enormous scale of the Morning Glory cloud.

that runs up the River Severn, in Britain. 'It is almost certainly the
result of a collision of opposing sea-breeze currents over Cape
York,' Christie continued. 'But I'm not sure we really understand
the details of these disturbances – there are a lot of puzzling
features. Why, for instance, you get such a variety of Glories: some
just one or two solitary waves, others an extensive succession of
them; some propagating over huge distances, others hardly at all.'

Mindful of Ralph Abercromby's efforts to establish that clouds
are much the same the world over, I asked Christie if the same sort
of cloud doesn't appear elsewhere. 'It occurs in the central United
States,' he answered. 'There's an example in the English Channel
and the Berlin fog waves, most notably those of 1968, are very
similar. It occurs in Eastern Russia and it's been seen in virtually all
the maritime areas of Australia.'

What? Was he telling me that I had come to the other side of
the world to see a cloud that I could have found in the bloody
English Channel? He reassured me that Burketown was worthy of
my cloudspotting voyage, for its Morning Glory is of a different
magnitude to other known examples and – what was more
important for me – 'it is the only known place, even after all these

years, where at a certain time of year there is a high probability you'll see one.' It is just not predictable anywhere else.

Asking Christie about the climatic conditions I should be looking out for, he told me to wait for 'a good sea breeze, blowing throughout the day, which will tend to carry a lot of water vapour as well as set up a good "waveguide"'. This, I learn, is like a smoothing of the air along the route that the cloud travels, so that the solitary wave is less likely to break up and dissipate as it approaches the coast. When these conditions coincide with a ridge of high pressure over the Cape York Peninsula, Christie told me, 'you're as good as guaranteed to see a Morning Glory.'

The locals aren't the type to care much for 'ridges of high pressure'. They have their own, less scientific, methods of anticipating the cloud's arrival. One, not surprisingly, involves beer: the glass doors to the pub's fridges are supposed to frost up when there is enough moisture in the air for the cloud to form. The other is that the cheap wooden tables in Paul and Amanda's café can be seen to bend upwards at the corners for the same reason.

But after two more mornings of rising at 4.30 just in case, the fridge doors were clear and the café tables as flat as the savannah, which stretched off endlessly in all directions.

☁

I WAS NOT THE ONLY ONE fretting. Ken Jelleff, a 48-year-old timber merchant from near Melbourne, had spent twelve days driving the 1,800 miles to Burketown in a 4WD with his wife and his microlight. We were having a 'mugaccino' at the distinctly level tables of the café after yet another cloud-free morning.

This was Jelleff's fifth pilgrimage to Burketown to soar the Morning Glory. He'd always been lucky enough to encounter one in past years, but this time he was beginning to wonder if his luck had run out. 'We've been here for five days now,' he told me, 'and we've got to head back in a day or so.'

Jelleff remembered a pilot from a few years back, who came for a fortnight and didn't see a single Morning Glory. He stayed an extra week, and then one more, but it still refused to show, only for one to arrive the day after he left. 'If our luck doesn't change,' he said, with a nervous smile, 'the same could happen to us.'

With such huge distances to tow their aircraft and the distinct possibility of no cloud arriving, is it really worth all the bother? 'You bet,' Jelleff enthused. 'When you're on the Glory, you're surfing a cloud and the wave of air causing it is as clear as crystal to fly through. It's unparalleled – the ultimate gliding experience.'

To stay up, a glider needs lift, or rising air, and this is something that a Morning Glory has in abundance. Invisible, solitary waves of air occur frequently around the world, but only in this region are the conditions right for roll clouds to form at their centres. The Morning Glories show the gliders where the waves are. They are beacons that render the atmosphere's movements visible.

Unfortunately for the Jelleffs, two more days of heading out to the airstrip at daybreak led to nothing. They could wait no longer, and so had to pack up the microlight and start the long drive back to Melbourne in the morning. 'There's always next year,' Ken said, trying to sound cheerful.

Just under a week into my stay, I wasn't feeling as philosophical about having to face the same fate. An old Aboriginal lady, living on nearby Bentinck Island, was said to know a traditional dance that brings the wind which carries the Morning Glory. Without hesitation, I arranged to go there in search of spiritual assistance. It seemed a long shot but, frankly, I was getting desperate.

THE LADIES OF BENTINCK were chatting in the shade of some mangrove trees, as my boat dropped me on the island's concrete jetty. 'Sunday we have break day,' Netta Loogatha explained when I got talking to them, 'so we don't go fishing today.'

Their grandchildren, visiting for the day, were playing in the surf. Aside from these visits, the ladies were alone on Bentinck. No men had lived there since the ladies banned alcohol on account of the problems it caused amongst the depressed Aboriginal communities of the region.

It soon became clear that the Aboriginal ladies had a very

different relationship with the Morning Glory from that of the visiting glider pilots and scientists. 'We used to call it the "yippipee",' Netta told me. 'That was the language name for it.' The name has a meaning: that the cloud brings the wet season, which starts around late October.

Did Netta remember the Morning Glory when she was young? 'My mother used to say that when the wind comes, it brings the yippipee and you need to look after your little brother and get ready. Then the waterspouts would come – we'd call that "dunderman" – it's very dangerous.'

And, as if to underline this point, she told me how a small plane had crashed into the sea some years before, while taking four of their community from the next island over to the mainland. The crash was caused by a freak occurrence of several Glories appearing at the same time. 'From every direction the yippipee just came – all of a sudden,' she explained, as the other ladies looked to the sand. The crash had happened in the early hours of the morning, I learnt. 'You could see them clouds rolling, you know, meeting each other, and the boys were laughing saying, what's wrong with them flying in a weather like this?'

Netta's voice wavered as she told me that her sister had been on the plane. 'In the afternoon we just saw Paulie Poole's plane searching. There were no bodies… they just found my sister's bag.' The anniversary of their deaths had been the previous day.

I remembered that Doug Christie had explained that Morning Glories don't just come from the usual northeasterly direction of the Cape York Peninsula. More rarely, they arrive from the south and the southeast too. In these instances, he suggested that the wave of air forming the cloud originates from thunderstorm activity over the highlands north of Mount Isa. I had also heard that when Morning Glories cross from different directions the air at their intersection was extremely turbulent and dangerous.

In spite of the tragedy for their small community, the ladies of Bentinck Island seemed to hold no grudges towards the cloud. They appeared resigned to the destructive power of Nature as something beyond their control. Netta even offered to ask Dawn Naranachil, one of the oldest amongst them – who didn't speak a

Dawn dances to bring the Glory.

word of English – to perform her 'wamur' dance to bring the wind that carries the Morning Glory.

I watched as Dawn lifted a handful of sand from the beach and threw it into the air, after which she began stamping her feet and chanting. She beckoned the grandchildren up from the surf to join in with her. They giggled as they copied her movements, and it felt as if a ritual was being passed down. 'You feel that wind come?' asked Netta.

A light breeze did seem to pick up but – to tell the truth – I was sceptical that the faint rustling of the palms as I departed would ever be enough to bring the yippipee.

☁

AND SURE ENOUGH, nothing came the following morning.

'We're not God,' said Paul Poole, eyeing my despondent expression over breakfast. 'We can't just turn them on and off when we want – bloody Pommie tourist bastards want everything your own way!'

Throughout the day, however, I got the feeling my luck was changing. A sea breeze picked up from the northeast and, in the afternoon, the barman in the pub drew my attention to a faint frosting on his fridges. Looking out of the window at 5am the following morning, I could see a dark line on the horizon. This was it.

I fumbled into my clothes and rushed out on to the empty road in front of the cabin. It was still dark, and a dog was barking manically. I could feel the wind picking up around me as the cloud reached the end of the street.

The light from the full Moon lent a silky, glacial sheen to the towering front of the cloud, which stretched off to the horizon in both directions. I stood transfixed as it made its way down the street towards me at a little over the town's speed limit. Ripples and undulations on its front face rose up and disappeared over the top as it progressed, making it appear to revolve back on itself. It was immense – blocking out the Moon and the Southern Cross as it passed over, casting the town into shadow. The back of the roll looked quite different from the front – a falling, Cumulus-like wall of cauliflower mounds, silver and black in the moonlight.

This was the cloud that I had crossed the world to see. Yet it had arrived before dawn and so had been only partially visible. Heading back to my cabin for a cup of coffee, it felt as if I'd come to hunt an infamous shark and had just caught a first glimpse of its fin breaking the surface of the water. I couldn't wait to see what this beast would look like in the full light of day.

☁

I FOUND MORE GLIDERS at the airstrip. Rick Bowie had come the 1,200 miles from Byron Bay, where he operates joy flights for the local gliding club. This was his third year in Burketown and he'd brought a self-launching Pik 20E glider with him. The plane has a two-stroke engine on the end of a retractable arm. Bowie explained that he can use the engine to get up in the air and then wind it away into a hatch in the plane's fuselage as he starts to glide. 'On a Glory with good lift,' Bowie explained, 'you can really go in this thing. I've been at speeds of 160mph.'

The dependable lift at the front of the cloud provides ideal conditions to perform the most adventurous gliding manoeuvres: 'You come over the face of the cloud and surf down it.' Bowie was showing the gliding movements with his outspread hand. 'You can have a wing tip in the cloud and go along the face – right down to the bottom edge of it. You can do aerobatics, loops…' But isn't it dangerous surfing the Morning Glory? 'Yeah, you have to treat this rolling wave with a lot of caution. You just don't know what the cloud's going to do – the lift can disappear as the wave comes inland from the sea and fizzles out.'

The turbulent, sinking air at the centre and rear of the cloud is to be avoided at all costs. When surfers 'wipe-out' on the average ocean wave, they get wet. If a glider pilot does the same at 4,000ft, while surfing the Morning Glory, the result is rather more drastic: 'It's isolated country out here – full of crocodiles,' warned Bowie. 'If you gotta use your parachute, no one's coming to get you in a hurry.'

In spite of its dangers – or perhaps because of them – the Morning Glory offers pilots the opportunity to break both distance and speed records. And this is something that Dave Jansen, a Qantas pilot from Mooloolaba, north of Brisbane, was in a position to talk about. In Burketown for the first time, Jansen knew as much as anyone about breaking records: not only had he been Australian national gliding champion five times, he was also the Records Officer for the Gliding Federation. He was responsible for verifying any national record attempt claims.

'The "Free Three Turn Points" distance record is the one that you could beat here,' he told me. Ah yes, the old three-pointer – I knew it would be that one… 'It is a flight that would take about eleven hours, and it's just a matter of time before someone does it on the Glory.'

Jansen was there with three other gliders, based at Lake Keepit Soaring Club in New South Wales. They'd brought an ASH 25 glider with them, which looked like a very fancy piece of kit, and boasted an enormous 87ft wingspan. Why, if records could be broken on the Morning Glory, were there no more than a handful of pilots here to surf it? His answer was instant: 'It's a bloody long way to come. There's nothing here. It's hot, it's isolated and there are millions of flies.'

By contrast to Jansen, Geoff Pratt seemed to be oblivious to the hardships of coming to Burketown. A gently spoken electrician from Cairns, he had arrived the previous evening after a fifteen-hour drive towing his Monerai V-Tail self-launching glider. He had made the same journey for the last nine years. Did he not get bored

ABOVE: The fridges in the Burketown Pub are frosting over and the café tables are curling upwards – both are signs that the Morning Glory is coming.
RIGHT: Geoff Pratt prepares to take off in his Monerai self-launching glider.

through the long, hot days just waiting for the cloud to show up? 'Burketown's not such a bad place to hang out,' he reflected, his gold tooth glinting in the morning sunshine. 'I like that it is so isolated. Word hasn't really spread yet and maybe that's a good thing. I wouldn't want it to become like a traffic jam.'

That evening, there was a palpable mood of anticipation amongst the pilots. All signs pointed towards another Glory appearing in the morning: the sea breeze had kept up all day, the café tables were dramatically bending up at the corners and, as we stood at the bar of the Burketown Pub, Frankie Wylie and I concluded that there was an indisputable frosting of the fridge doors. It was surely a dead cert.

The only question in my mind was whether the cloud would arrive at a less ungodly hour. This was not because such early mornings were hard work, but because the gliders would not be able to go up and surf the Morning Glory unless it was considerate enough to arrive in daylight.

☁

ALL THE USUAL SUSPECTS were on the airstrip by the time I arrived at 5am. I helped Rick Bowie wipe the dew from the wings of his Pik 20E. Whilst this was a great sign of enough moisture in the air for the cloud to form, he explained, the dew changed the flow of air over the wings, making the glider more difficult to control. Paul Poole had come to the airstrip too, agreeing to take

Both images: Russell White (member 23)

me up in his Cessna 187 motor-plane so I could see the cloud close-up. 'There's definitely one out there,' he said, peering towards the dawn horizon. As soon as it was light enough to fly, the pilots rushed to their planes, and took off in quick succession towards the rising Sun. Paul and I were in hot pursuit.

Twenty miles north of the town we reached the coast. And there, rolling in towards us, was not one Morning Glory cloud but three. The front one had a silky smooth surface, making it look like an enormous snowy glacier, suspended 500ft above the ground. The second and third clouds were rough and puffy, propagating in the turbulent wake of the first.

From the air, I could see the enormous length of the clouds, which snaked out in both directions along the Gulf. Where a section of the front wave of air had passed over Bentinck Island, its progress had been slowed, compared with the parts over the open water around, and this gave the line of the cloud a distinct kink. Before take-off we had removed the Cessna's side door from its hinges, so that there was no glass between me and the cloud. It looked so clean and smooth and bright. I wanted to jump out to it.

The glider planes looked tiny against the primary cloud. Just like surfers on the biggest swell that Hawaii's Waimea Beach could muster, the pilots soared down its leading edge. They picked up speed as if taking their gliders into a dive but, in the continual lift at the front of the wave, this was a dive with no loss of altitude. Then they climbed the gradient of the cloud face, dropped a wing and banked in a steep curve to head back in the other direction.

Off in the distance, I saw Rick Bowie looping the loop ahead of the face. Geoff Pratt gained altitude and jumped over from the 'primary' to ride the rougher second and third clouds. The glider wings glinted, like white, waxed surfboards in the low morning Sun, as they tore along the immense swell of the atmosphere.

I thought how furious Ken Jelleff would have been to see what he'd missed this year. 'Ten or fifteen minutes into the flight,' he had told me, 'the Sun comes over the top of the cloud. When you look back at that enormous fluffy wave, with the golden Sun breaking behind it, it looks like something the Italians would have painted in the Renaissance. You'd swear you were in heaven. It's that good.'

Both images: Russell White (member 23)

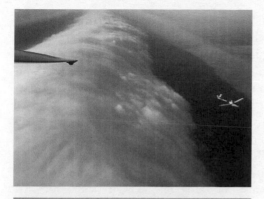

Below: Russell White (member 23)

Geoff Pratt cruises along the front of the wave
of air travelling around the cloud.

And Jelleff was right. The cloud was resplendent. 'Until you actually experience what it is like to soar this cloud,' Geoff Pratt had been enthusing only the day before, 'you don't understand how special it is. Sometimes I have to pinch myself, just to know that I am actually doing it.'

I had crossed the world to meet this cloud and, finally, here we were, face to face. I held my hand to shield my eyes from the brilliant rays, now that the Sun was well off the northeastern horizon. And these cascaded down the cloud's face, casting long, warm shadows along the ripples of its surface. The undulations gently rose up with the progress of the wave, before disappearing over the crest.

WHAT MUST IT HAVE BEEN like to be the first person to do this – to have gone up and surfed this cloud before anyone else – before others had even determined that it was possible? Russell White is one of only two people in the world who are in a position to say. The pilots in Burketown had often mentioned his name – clearly considering him a gliding legend – for the whole craze of soaring the Morning Glory owes its existence to a pioneering flight White made with his gliding partner, Rob Thompson, in the spring of 1989. Though a regular face amongst the Morning Glory chasers, this time White had been unable to make it to Burketown. However, I spoke to him on the phone at his home in Byron Bay, to ask about their first encounter with the cloud.

White and Thompson had flown up to the Great Barrier Reef to spend a few days sailing when the skipper told them about this extraordinary cloud. Both pilots had for many years glided on the air streams around mountains, which give rise to the Altocumulus

The first Glory hunters lenticularis clouds that look like UFOs. The stationary, standing waves of air associated with these types of cloud are different from the moving waves that cause the Morning Glory. The difference is comparable to that between the fixed crest on the surface of a stream as it passes over a boulder and the travelling waves on a beach, but both men felt convinced that the gliding principles would be the same for both air streams. They felt sure that the moving Morning Glory wave had the potential for a fantastic gliding experience, and over a beer on the boat they resolved to fly to Burketown the following day in White's motor-glider to see if they could find it.

'We arrived late in the afternoon of October 12,' he remembered, 'but no sooner had we found a place to stay than I was told of a disaster at work, which meant I'd have to go back the next day.' Despondent, they went to bed, deciding that, if by some miracle, it arrived in the morning, they'd attempt to fly it before departing.

'The alarm clock had been left in the plane and I was still in the shower when Rob came racing into the cabin, shouting "It's coming!"' In a desperate rush, they hitched a lift out to the airfield, and as they taxied down the airstrip they were almost underneath the advancing cloud. 'We took off away from the Glory,' White told me. It is now established that gliders should never take off towards the cloud. 'At that time, there were no rules – we were making them up as we went along.'

Turning their glider towards the cloud, they felt the lifting air at well below 1,000ft. 'It was the most remarkable flight,' White raved. 'It left us awestruck and ecstatic – completely over the moon that

Russell White (member 23)

Russell White, the Morning Glory pioneer.

we'd found this thing and were surfing it. It was just outstanding.' The Morning Glory they surfed that momentous day was a relatively modest one – only some thirty miles long and 3,000ft high.

The pioneering hour-and-a-half flight left them hooked, however. On their way back south, they stopped off at Lake Keepit in New South Wales, home to one of the largest gliding clubs in the country, and announced to all and sundry that they'd just flown the Morning Glory. 'They didn't believe us,' White laughed. 'Seriously, they didn't believe us – they thought we were making it all up. So we went back the following year with cameras.'

Through White's articles in the gliding press, and a short documentary subsequently filmed by Thompson, word spread and others began to make the spring pilgrimage to Burketown in search of thrills. To this day, however, White estimates that those who have actually flown the cloud number only a few dozen. Was he proud of what he began? 'I'm pleased as punch. Could you describe the Himalayas to somebody who hasn't seen them and do them justice? No way – you have to see them for yourself. It's the same with the Morning Glory: it's such an awesome experience but you just have to be there to get it.'

☁

BACK AT PAUL AND AMANDA'S, over a celebratory dinner of Barramundi fish, washed down with the amber nectar, I explained to the collected pilots that I had recently founded The Cloud Appreciation Society. Like a B-list actor promoting his latest film, I launched into a well-rehearsed speech in defence of our fluffy friends. Life would be dull, I declared, had we nothing but blue monotony to look at, day after day. I mentioned how Ralph Waldo Emerson, the American essayist, described the sky as 'the daily bread of the eyes... the ultimate art gallery above.'[4] And that the society therefore stands in opposition to 'blue-sky thinking' wherever we find it.

Clouds are the face of the atmosphere, I proclaimed, enthusing on their ability to express its moods and communicate the invisible architecture of its currents. And then – as I was moving on to the

Russell White (member 23)

part about the clouds being Nature's poetry – I caught the glint, once again, of Geoff Pratt's gold tooth.

Along with the other pilots, he was smiling at me.

What a fool I was. Had I not been listening when he told me of his own relationship with the clouds? 'When I'm flying amongst them,' he had confided out at the airstrip, 'I feel like I'm at home. Up there, I'm with the soaring birds – birds like the Wedge Tail Eagle – and they let me fly with them. Up in the clouds, you can't help but have a belief in a creator.'

Who, other than fellow cloudspotters, would have made the journey to such a remote, offbeat little town? I had crossed the world only to find, as I have so many times since, that I was preaching to the converted.

Notes

1. CUMULUS

1 Descartes, René: from Vrooman, J.R., *René Descartes: A Biography*, 1990.
2 Ruskin, John: *Modern Painters*, Volume III (1856), Chapter XVI.
3 Ovidius Naso, P.: *Metamorphoses*, ed. Brookes More.
4 Anon, *A Book Of Contemplation The Which Is Called The Cloud Of Unknowing, In The Which A Soul Is Oned With God*, Chapter 3.
5 Ibid., Chapter 9.

2. CUMULONIMBUS

1 This, and all other Rankin quotes, come from *The Man Who Rode the Thunder* by William H. Rankin, 1960.
2 The Luke Howard references throughout this book draw heavily on Richard Hamblyn's fine biography, *The Invention of Clouds: How an Amateur Meteorologist Forged the Language of the Skies*.
3 Shakespeare, William: *King Lear*, Act III, Scene II.
4 Hale, M.R. trans.: 'The Meghaduta of Kalidasa', stanzas 7 & 8.
5 Ibid., stanza 24.

3. STRATUS

1 Sandburg, Carl: 'Fog', *Chicago Poems*.
2 Hugo, Victor: *Victor Hugo's Intellectual Autobiography*, trans. Lorenzo O'Rourke.
3 Stieglitz, Alfred: Letter to J. Dudley Johnston, 3 April 1925.
4 Stieglitz, Alfred: Letter to Heinrich Kühn, 14 October 1912.
5 Stieglitz, Alfred: Letter to Hart Crane, 10 December 1923.
6 Stieglitz, Alfred: 'How I Came to Photograph Clouds', *Amateur Photographer and Photography*, 19 September 1923.
7 Lowell, James Russell: *The Vision of Sir Launfal*.

4. STRATOCUMULUS

1 Aristophanes: *Birds*, Loeb Classical Library, ed. & trans. Jeffrey Henderson.
2 Thoreau, Henry David: Journal, 10 July 1851.
3 Swift, Jonathan: *Travels into Several Remote Nations of the World by Lemuel Gulliver*, Part III, Chapter II.
4 Ibid.

5. ALTOCUMULUS

1 Shakespeare, William: *Hamlet, Prince of Denmark*, Act III, Scene II.
2 Aristophanes: *Clouds*, Loeb Classical Library, ed. & trans. Jeffrey Henderson.
3 Lucretius: *De Rerum Natura*, IV. Loeb Classical Library, trans. W.H.D. Rouse.
4 Wilde, Oscar: *The Decay of Lying: An Observation*.
5 Aristotle: *On Dreams*, trans. J.E. Beare.

6. ALTOSTRATUS

1 Thoreau, Henry David: Journal, 7 September 1851.
2 Thoreau, Henry David: Journal, 20 July 1852.
3 Thoreau, Henry David: *A Week on the Corcord and Merrimack Rivers*.
4 *King James Bible*, Matthew, Chapter XVI.
5 Thoreau, Henry David: Journal, 25 December 1851.
6 Keats, John: *Lamia*, Part II.

7. NIMBOSTRATUS

1 Milton, John: *Paradise Lost: The Second Book*.
2 Keats, John: 'Ode on Melancholy'.
3 Shelley, Percy Bysshe: 'Adonais'.
4 Updike, John: *Self Consciousness: Memoirs*.

8. CIRRUS

1 Howard, Luke: *On the Modifications of Clouds, and on the Principles of Their Production, Suspension, and Destruction … Philosophical Magazine* XVI (1802).
2 Yeh Meng-te: from *The Weather Companion* by Gary Lockheart.
3 Pliny the Elder: *The Natural History*. Book II, ed. John Bostock & H.T. Riley.
4 Varahamihira: *Brihat Samhita*, Chapter 32, trans. M. Ramakrishna Bhatt.
5 Li, D.J.: *Earthquake Clouds*. Xue Lin Public Store, Shanghai, China.
6 Shou, Zhonghao & Harrington, David: 'Bam Earthquake Prediction and Space Technology', presented at United Nations/Islamic Republic of Iran Regional Workshop on the Use of Space Technology for Environmental Security, Disaster Rehabilitation and Sustainable Development, Tehran, Islamic Republic of Iran, 8–12 May 2004, available from the UN Office for Outer Space Affairs website (www.oosa.unvienna.org).

10. CIRROSTRATUS

1 This interpretation was brought to my attention by Gary Lockhart's book, *The Weather Companion*.
2 www.meteoros.de/indexe.htm
3 www.forspaciousskies.com
4 Whilst there are some instances of the *labarum* symbol appearing prior to the Battle of Milvian Bridge, these are extremely rare and are not necessarily employed in a Christian context.

12. CONTRAILS

1 Venkataramani, M.S.: 'To Own The Weather', published in *Frontline*, India's National Magazine, 16–29 January 1999.
2 Hersh, Seymour M.: 'Rainmaking Is Used As Weapon by U.S.', published in the *New York Times*, 3 July 1972.
3 Frisby, E.M.: 'Weather Modification in Southeast Asia, 1966–1972', *The Journal of Weather Modification*, April 1982.
4 Breuer, G.: *Weather Modification*.
5 Congress, 93rd, 2nd Session, 1974. Hearing, 20 March 1974. 'Briefing on Department of Defense Weather Modification Activity' by Col. Soyster.
6 Ibid.
7 House, Col. Tamzy J.; Near, Lt-Col. James B. Jr.; Shields, L.T.C. William B. (USA); Celentano, Maj. Ronald J.; Husband, Maj. David M.; Mercer, Maj. Ann E.; Pugh, Maj. James E.: *Weather as a Force Multiplier: Owning the Weather in 2025, A Research Paper Presented to Air Force 2025*, August 1996.
8 Venkataramani, M.S.: 'To Own The Weather', published in *Frontline*, India's National Magazine, 16–29 January 1999.
9 Changnon, S.A.: *Natural Hazards of North America*.
10 *IPCC Special Report on Aviation and the Global Atmosphere, 1999*.
11 Minnis et al, 2002: 'Spreading of isolated contrails during 2001 air traffic shutdown'. American Meteorological Society, J9–J12.
12 Travis, D.J.; Carleton, A.M.; Lauritsen, R.G.: 'Contrails Reduce Daily Temperature Range'. *Nature*, 8 August 2002.
13 Minnis, P.; Ayers, J.K.; Palikonda, R.; Phan, D.: 'Contrails, Cirrus Trends, and Climate'. 2004, *Journal of Climate*, 17.
14 Mannstein, H. & Schumann, U.: 'Observations of contrails and Cirrus over Europe'. Proceedings of the AAC Conference, 30 June–3 July 2003, Friedrichshafen, Germany.
15 *IPCC Special Report on Aviation and the Global Atmosphere, 1999*.
16 Williams, V. & Noland, R.B.: 'Variability of contrail formation conditions and the implications for policies to reduce the climate impacts of aviation', not yet published.
17 Brooke, Rupert: 'Clouds'.

13. THE MORNING GLORY

1 Scorer, Richard & Verkaile, Arjen: *Spacious Skies: The Ultimate Cloud Book*.
2 Abercromby, The Hon. Ralph: *Seas and Skies in Many Latitudes, or Wanderings in Search of Weather*, 1888.
3 Ibid., page vii.
4 Emerson, Ralph Waldo: Journal, 25 May 1843.

Picture acknowledgements

All cloud photographs are reproduced by kind permission of The Cloud Appreciation Society members. Photography credits can be found on the relevant pages. All photographers retain copyright for their images. Where there is no credit on the page, the photograph is © Gavin Pretor-Pinney.

14–15: Anthony Haythornthwaite. *25 (above):* Jasmine De Aragüés. *36:* Kunsthistorisches Museum, Vienna/photo Bridgeman Art Library. *48, 69:* Taken from *The Man Who Rode the Thunder* by William H. Rankin, Prentice-Hall, Inc., Englewood Cliffs, NJ, 1960. 67: Image reproduced with permission of Walter Lyons, © Sky Fire Productions, www.Sky-Fire.TV. *81 (both images):* © Beat Widmer, design by Diller Scofidio + Renfro. *83:* Debra Hill Prods/The Kobal Collection. *100:* © Nicolas Reeves & NXI GESTATIO. *105:* Olana State Historic Site, Hudson, NY. New York State Office of Parks, Recreation and Historic Preservation. *117:* San Francesco, Arezzo/photo Bridgeman Art Library. *124:* Kunsthistorisches Museum, Vienna/photo Bridgeman Art Library. *157:* Getty Images. *173:* J.C. Dollman, from *Myths of the Norsemen* by H.A. Guerber, George G. Harrap & Co Ltd, London, 1922. *191:* IndoEx satellite image from 21 December 2003. *202:* Eric Winter, from *The Princess and the Pea* © Ladybird Books, Ltd., 1967. Reproduced with permission of Ladybird Books, Ltd. *209 (fish illustrations):* Anthony Haythornthwaite. *235:* British Museum (1890.8.4.11). *261, 263:* Schenectady Museum, Schenectady, NY. *265:* © The New York Times Company, 1972. Reprinted with permission. *290:* Courtesy of National Oceanic & Atmospheric Administration USA (NOAA).

Quotation acknowledgements

Permission to use copyright material is gratefully acknowledged. The Alfred Stieglitz quotations on p87, taken from *Alfred Stieglitz: Photographs and Writings*, edited by Sarah Greenough and Juan Hamilton, National Gallery of Art, 1998, are © 2006 The Georgia O'Keeffe Foundation/Artists Rights Society (ARS), New York. The quotation on p167, from John Updike's *Self Consciousness: Memoirs*, 1989, is by permission of John Updike. The quotations in Chapter 2 from William H. Rankin reprinted with the permission of Simon & Schuster Adult Publishing Group, from *The Man Who Rode the Thunder* by William H. Rankin. Copyright © 1960 by Prentice-Hall, Inc; copyright renewed © 1988 by Prentice-Hall, Inc. Whilst every effort has been made to trace all copyright holders, the publishers will be pleased to rectify any omissions in future editions of the book.

Author's acknowledgements

Thanks very much to the following for their help with this book:
Richard Atkinson, my editor at Hodder, for showing an interest
in it at a stage when every other publisher in London was
uninterested. Also, for all the hard work that he put into it with
me, and for being open-minded enough to allow me to design it.
Patrick Walsh, my agent, for his invaluable help improving the
book proposals, and for providing support and continuity
throughout. Ross Reynolds, at the Department of Meteorology,
University of Reading, for providing his precious time to
read the manuscripts, and clarify and correct the science parts.
Roderick Jackson, without whom I'd never have had the
confidence to consider writing it.
Thanks also to:
Cathy and Peregrine St Germans and Simon Prosser, at the Port
Eliot Literary Festival, for inviting me to give the cloud talk
during which The Cloud Appreciation Society was founded.
Michele Lavery, at the *Daily Telegraph*, for commissioning me to
write about the Morning Glory cloud – Chapter 13 is based on
the article subsequently published in the *Telegraph*'s Saturday
magazine. Qantas Airlines, for flying me to Australia. Nicola
Doherty, Henry Jeffreys, Juliet Brightmore, Auriol Bishop and
Jocasta Brownlee at Hodder & Stoughton. Steve Dobell, for his
fine copy editing. All the friends I made on sabbatical in Rome,
who provided much inspiration. Philip Eden, for his advice at the
beginning. Tim Garratt, for his advice at the end. Jack Borden,
of For Spacious Skies in the USA, for encouragement and a steady
supply of quotes and leads. Claire Paterson, for much-valued
impartial advice.
And finally:
Liz Pickering, my wife-to-be. Alex Bellos, for reminding me at
early stages to get to the point in my writing. Tom Shone and
Tom Hodgkinson, for reassuring me, many years ago, that a book
about clouds wasn't a stupid idea.
All the members of The Cloud Appreciation Society who have
contributed their fantastic photographs.

Index